ENVIRONMENTAL REMEDIATION TECHNOLOGIES, REGULATIONS AND SAFETY

OFFSHORE WIND

TECHNOLOGIES, ECOLOGICAL RISKS AND PROSPECTS

Environmental Remediation Technologies, Regulations and Safety

Additional books in this series can be found on Nova's website under the Series tab.

Additional e-books in this series can be found on Nova's website under the e-book tab.

ENVIRONMENTAL REMEDIATION TECHNOLOGIES, REGULATIONS AND SAFETY

OFFSHORE WIND

TECHNOLOGIES, ECOLOGICAL RISKS AND PROSPECTS

CHESTER MENDOZA
EDITOR

nova publishers
New York

Library of Congress Cataloging-in-Publication Data

Offshore wind (Mendoza)
Offshore wind : technologies, ecological risks, and prospects / editors, Chester Mendoza.
pages cm. -- (Environmental remediation technologies, regulations and safety)
Includes index.
ISBN 978-1-63482-364-7 (hardcover)
1. Offshore wind power plants. 2. Wind power industry. I. Mendoza, Chester. II. Title.
TK1541.O33 2015
333.9'209162--dc23

2015006804

Published by Nova Science Publishers, Inc. † New York

CONTENTS

PREFACE

Offshore wind has been envisaged as making a critical contribution in Europe in satisfying the electricity demand in a low-carbon generation scenario. In fact, offshore environment provides a windier and steadier source compared to onshore wind energy. Offshore wind turbine rotors and generators create value by "farming" the wind, converting its energy into electricity. Costly infrastructure – the very high towers, deep seabed foundations, buried high voltage insulated cables, on-platform transformers and complete subsea transmission grid – and expensive at-sea maintenance are required because this electricity is not where it is needed. This book discusses offshore wind technologies and examines the ecological risks and prospects of offshore wind.

Chapter 1 - Offshore wind has been envisaged as making a critical contribution in Europe in satisfying the electricity demand in a low-carbon generation scenario. In fact, offshore environment provides a windier and steadier source compared to onshore wind energy.

Installation, operation and maintenance of offshore facilities, whether associated with the generation, collection or transmission of the electrical energy, is extremely expensive and so the correct technologies must be deployed to provide the maximum cost-benefit. For quite long distances offshore, High-Voltage Direct Current (HVDC) transmission is preferred for economic and technical reasons; hence, this technology provides the platform that can be used to enable massive integration of offshore wind farms into AC onshore networks with minimum losses and increased flexibility over power control.

In case of many, large offshore wind farms, conventional point-to-point connections to-shore appear inadequate for operational, market and reliability reasons. Multi-Terminal HVDC (MTDC) grids are thus foreseen to increase flexibility and to reduce the power not supplied to onshore grids in case of DC disturbances.

Voltage Source Converter (VSC) technology is the most promising converter solution for the development of offshore DC connections. Though less mature than Line Commutated Converter (LCC) technology, VSC offers significant advantages for offshore grid applications, moreover it is rapidly evolving.

It is envisaged that such grids will provide significant benefits beyond the only integration of multiple offshore wind farms dispersed over wide areas into AC onshore grid. A fully operational MTDC grid with offshore Wind Farms (WF) can be regarded as a large (virtual) power plant able to provide services to the mainland AC grids. In addition to optimisation of AC and DC transmission infrastructures, and potential improvement of reliability and security of supply, MTDC grids are expected to realise (1) wind power transfer function (including smoothing of wind power fluctuations), (2) interconnection function (i.e., use of the MTDC grids to exchange power between AC zones), (3) ancillary services provision (e.g., voltage support, frequency support to onshore AC grids). In spite of those potential benefits, no MTDC grids currently exist, as major barriers still remain.

This chapter presents some of the activities carried out within the research project TWENTIES (2010-2013), co-funded by the European Union, and subsequent developments carried out by partners, addressing technical solutions for MTDC grid design and control to achieve the above expected functionalities. The chapter starts by presenting the different functions an HVDC grid may fulfil. Subsequently, possible topologies of the future DC grids are discussed, emphasizing the motivations and advantages brought by them. After illustrating converter technology features, having impacts on key aspects such as protection and interoperability of the DC grid, the operation of a DC grid is introduced under normal and emergency conditions. DC protection issues are then discussed. Ancillary services provision will be finally presented.

Chapter 2 - North Sea-bordering countries lead the world in offshore wind farm development, with scientific and engineering breakthroughs, as well as pitfalls, being revealed through these countries' offshore wind turbines. The majority of UK and other North Sea offshore wind turbines use monopole foundations. Grouted connections have been used as industry standard for these turbines' foundation designs. The recent UK case *MT Højgaard a/s v. E.ON Climate and Renewables UK Robin Rigg East Limited*, et al. relates to failed grouted connections due to flawed offshore wind turbine foundation design technology in one of the UK's largest wind farms, the Robin Rigg offshore wind farm. Ruling on this case, the UK's High Court of Justice determined that following industry standards for scientific and engineering design is insufficient to shield designers/contractors from liability when a design fails to meet a contractually-mandated "fitness for purpose" requirement for each turbine to achieve a minimum 20 year service life.

While *MT Højgaard* may help to inform the development of the U.S.'s infant offshore wind industry, strict adherence to certain aspects of this international precedent may hinder such industry's effective launch. Most U.S. Wind Energy Areas for offshore wind development are located along the Mid-Atlantic East Coast. Based on recent events and scientific data, hurricanes now travel further up the East Coast than they have historically. Currently, offshore wind turbine blades are being laboratory-tested to improve their designs and increase their resiliency to hurricanes and extreme weather conditions. Nevertheless, such blades, while at the forefront of innovative engineering and scientific technologies, will remain untested in practice due to the U.S. currently having no offshore wind farms. Accordingly, such blades may fail to withstand actual, successive hurricanes over time. If such blades are contractually obligated to meet a minimum 20 year service life "fitness for purpose" performance standard, they may fail to do so.

This chapter analyzes grouting issues impacting North Sea offshore wind turbines with monopole foundations. It also discusses *MT Højgaard*, using it as a springboard for examining contractual obligations when weighed against industry standards. Additionally, it examines scientific data regarding U.S. hurricanes and explores why cutting-edge scientific know-how, combined with state-of-the art engineering technology, may be insufficient to meet legal standards that are aspirational and potentially unrealistic. Finally, it discusses why the ruling and certain facts in *MT Højgaard* may hinder U.S. offshore wind development if the lessons learned from such case are not adopted.

Chapter 3 - Offshore wind turbine rotors and generators create value by 'farming' the wind, converting its energy into electricity. Costly infrastructure – the very high towers, deep seabed foundations, buried high voltage insulated cables, on-platform transformers and complete subsea transmission grid – and expensive at-sea maintenance are required because this electricity is not where it is needed. The cost of this infrastructure cripples the investment payback ratio. High investment across the world in specialised manufacturing plant and tens of thousands of experienced workers enable the rhythm and flow that mass produces some 20,000 wind turbines a year of the now-standardised designs of wind turbines erected on land close to where the energy is needed, driving their cost effectiveness. Offshore represents a 2.5% extension of the market, in disjointed projects, requiring the expensive development of much larger wind turbines and requiring yet more investment in much larger factories but offering no chance of continuous rhythm-and-flow production and also requiring the development of much new and expensive offshore infrastructure-building technology to make it possible at all. This kind of 'bigger' is not 'better'. Technical heroics is no substitute for the lean and agile thinking within the industry itself that is steadily developing lighter, larger, more dynamically responsive rotors and lighter, taller towers, enabling today's standardised wind turbines to cost effectively 'farm' the lighter winds close to where the electricity is needed, with minimal infrastructure costs.

Chapter 4 - This chapter presents a new method to assess techno-economically innovative solutions for offshore wind projects. The main aim of the proposed technique is to provide academic and industrial technologists with a new tool to conduct accurate cost benefit analyses in an agile way. The method proposes a simplified life-cycle economic analysis, taking into account the time value of money and the most relevant economic variables. Its structure also enables to study the effect of long term variations in wind speed, electricity price and other variables in conjunction in a simple and effective way. The assessment of the energy yield of the offshore wind power plant (OWPP) considers the electric power losses in all the systems that make up the windfarm, as well as the wake effect and the unavailability of the different systems. The formulation of the method enables it to be shaped either in a spreadsheet or in algorithmic form, avoiding power flow simulation studies up to now used by utilities and consultants to assess the electric power losses in grid integration studies. By doing so, the proposed method decreases the time needed to conduct the studies and increases staff capable of performing the task.

In: Offshore Wind
Editor: Chester Mendoza
ISBN: 978-1-63482-364-7

Chapter 1

HVDC Grid for Offshore Wind Exploitation[1]

***Emanuele Ciapessoni*[1], *Diego Cirio*[1*], *Andrea Pitto*[1], *Keith Bell*[2], *Grain Adam*[2], *Steven Finney*[2], *Chen-Ching Liu*[3], *Carlos Moreira*[4], *Olivier Despouys*[5] *and Antonio Gatti*[1]**

[1]RSE S.p.A., Italy
[2]Strathclyde University, UK
[3]Formerly at University College Dublin, Ireland
[4]INESC Porto, Portugal
[5]RTE, France

Abstract

Offshore wind has been envisaged as making a critical contribution in Europe in satisfying the electricity demand in a low-carbon generation scenario. In fact, offshore environment provides a windier and steadier source compared to onshore wind energy.

Installation, operation and maintenance of offshore facilities, whether associated with the generation, collection or transmission of the electrical

---

[1] Authors gratefully acknowledge contribution of Dr. Silvia Canevese for chapter reviewing.

* email diego.cirio@rse-web.it

energy, is extremely expensive and so the correct technologies must be deployed to provide the maximum cost-benefit. For quite long distances offshore, High-Voltage Direct Current (HVDC) transmission is preferred for economic and technical reasons; hence, this technology provides the platform that can be used to enable massive integration of offshore wind farms into AC onshore networks with minimum losses and increased flexibility over power control.

In case of many, large offshore wind farms, conventional point-to-point connections to-shore appear inadequate for operational, market and reliability reasons. Multi-Terminal HVDC (MTDC) grids are thus foreseen to increase flexibility and to reduce the power not supplied to onshore grids in case of DC disturbances.

Voltage Source Converter (VSC) technology is the most promising converter solution for the development of offshore DC connections. Though less mature than Line Commutated Converter (LCC) technology, VSC offers significant advantages for offshore grid applications, moreover it is rapidly evolving.

It is envisaged that such grids will provide significant benefits beyond the only integration of multiple offshore wind farms dispersed over wide areas into AC onshore grid. A fully operational MTDC grid with offshore Wind Farms (WF) can be regarded as a large (virtual) power plant able to provide services to the mainland AC grids. In addition to optimisation of AC and DC transmission infrastructures, and potential improvement of reliability and security of supply, MTDC grids are expected to realise (1) wind power transfer function (including smoothing of wind power fluctuations), (2) interconnection function (i.e., use of the MTDC grids to exchange power between AC zones), (3) ancillary services provision (e.g., voltage support, frequency support to onshore AC grids). In spite of those potential benefits, no MTDC grids currently exist, as major barriers still remain.

This chapter presents some of the activities carried out within the research project TWENTIES (2010-2013), co-funded by the European Union, and subsequent developments carried out by partners, addressing technical solutions for MTDC grid design and control to achieve the above expected functionalities. The chapter starts by presenting the different functions an HVDC grid may fulfil. Subsequently, possible topologies of the future DC grids are discussed, emphasizing the motivations and advantages brought by them. After illustrating converter technology features, having impacts on key aspects such as protection and interoperability of the DC grid, the operation of a DC grid is introduced under normal and emergency conditions. DC protection issues are then discussed. Ancillary services provision will be finally presented.

1. INTRODUCTION

With the significant deployment of off-shore wind farms especially in Europe, in order to achieve the EU renewable targets up to 2050, off-shore wind promises to become one of the important energy sources in the near future [1]-[9]. Wind farms with a total capacity of thousands of megawatts are expected to be installed in European waters by the next decades[2]. Plans are currently advancing for the development of such large-scale wind farms in Swedish, Danish, German, Dutch, French, Belgian, British and Irish waters. Compared with on-shore wind farms, off-shore ones have generally higher wind speed and larger power generating capability, which can enhance the wind turbine efficiency by 5-6%.

At present, off-shore wind farms or clusters of them are connected to shores only through radial connections. Depending mainly on the transmission distance and required capacity, these radial connections are either Alternating Current (AC) or Direct Current (DC). A DC solution becomes mandatory in case of far and/or high-capacity offshore wind farms, e.g., the North Sea offshore wind parks in service and under construction near the coasts of Netherlands, Germany, and Denmark (Borwin1: 400 MW - 203 km; Borwin2: 800 MW - 200 km; DolWin1: 800 MW - 165 km; ElWin1: 576 MW - 130 km; SylWin1: 864 MW - 205 km; DolWin2: 900 MW - 135 km), see Tables 1 and 2.

In parallel to radial connections dedicated to wind generation, point-to-point submarine interconnectors have been realised or are being planned to facilitate power exchanges between countries, thus contributing to a single European Electricity market: some examples are given by IFA 2000 (2 GW), NorNed I (700 MW), East Link I (350 MW), SwePol (600 MW - 2000), BritNed (1GW - 2011), COBRA (700 MW – 2019), NorBalt (700 MW - 2015).

It must be considered that combined solutions for off-shore interconnection and wind generation transmission based on mainly three-legged multi-terminal High Voltage Direct Current (HVDC) are currently under investigation in some projects, such as Kriegers Flak (600 MW – 2020) [10], COBRA [11] and Caithness-Moray HVDC link (800÷1200 MW – 2018,

[2] Based on [2], about 230 GW (17% offshore WFs), 400 GW (37% offshore WFs) and 735 GW (61% offshore WFs) will be installed within EU respectively by 2020, 2030 and 2050.

with hub option) [12], laying the foundation for future off-shore grids [13]-[17].

In Europe only the North Seas Offshore Grid Infrastructure (NSCOGI) scheme at 2030 includes 19 projects where, on the whole, about 10000 km of new DC subsea cables should integrate up to 112 GW of offshore wind power [13]. In particular, the NSCOGI study emphasizes the role of multi-terminal, meshed DC grid design versus the radial option in terms of lower grid investment.

As for DC transmission, the future offshore grid can be realised adopting two types of HVDC technologies: Line Commutated Converters (LCCs) and Voltage Source Converters (VSCs). LCCs are more mature (very high voltage and power levels, up to ±800 kV DC and 7.2 GW respectively, low losses), but VSCs appear more suitable for the Offshore DC grid concept itself. Currently the maximum rated DC voltage of the VSC technology is ±640 kV DC and the maximum power transmitted is 1200 MW.

In spite of the potential benefits, however, major barriers to the development of actual HVDC grids still remain[3]. One of the technical issues is related to the realisation of effective DC protection systems, which are mandatory to guarantee that a fault does not imply either damage to the electrical infrastructure or service interruption. In this regard, the DC breaker, i.e., one of the key elements of complex DC grids, is not yet a mature component and has currently no full-scale installation in operation. Another critical issue is that the implementation of DC grids will require future specific market rules and regulatory frameworks within the vision of the single European Internal Energy Market (IEM) [18]-[19].

When looking at the objectives beyond 2020 for off-shore wind generation, which could grow to 400 GW in 2030 and 735 GW in 2050 according to EWEA scenarios [1]-[2], and taking into account the magnitude of investment costs for connecting such amounts of wind generation, a long-term solution for the connecting infrastructures has to be developed on a very large scale. The solution should be supported by a harmonized legal framework to meet optimized economic criteria (to be defined) and provide a secure and sustainable energy supply.

Outside Europe, two recent projects in China emphasize the general interest for what can be regarded as the first tree-like DC grids in the world, connecting wind generation via VSC HVDC converters: the three-terminal

[3] Currently, very few radial HVDC multi-terminal connections exist, and no meshed HVDC grids.

Nan'Ao and five-terminal Zoushan DC links have been under exploitation since December 2013 and June 2014 respectively. They illustrate the benefits of using VSC converters in a multi-terminal scheme: such devices, in fact, exhibit inherent flexibility in connecting intermittent power (from wind farms) to various AC systems (from weak islanded AC networks to the strong AC grid at the onshore point of connection), while providing controlled and bi-directional power flow.

Table 1. Examples of HVDC transmission systems using current source converter technology

Project	Rating	Application	Distance	Commissioning year
Neptune (USA)	660MW, mono-polar 550kV DC	Interconnection of two areas	82 km submarine cable + 23 km land cable	2007
Guizhou-Guangdong II (China)	3000MW, bipolar ±500kV DC	Interconnection of two areas	1225 km	2008
Rapid city DC tie (USA)	200MW, bipolar ±13kV double circuit	Asynchronous connection (using back-to-back configuration) provides voltage and frequency support	0	2003
The world longest transmission line (Brazil)	2×3150MW HVDC line and 800MW back-to-back converters, both at ±600kV	Asynchronous connection of two areas	2500 km	2012
Chandrapur	2×500MW 205kV DC	Asynchronous connection using back-to-back HVDC link, provides frequency and voltage support	0	1997
Cross Channel HVDC between UK and France	2×1000MW ±270kV DC link	Interconnection of two areas as power trading facility	73 km	1986

Table 2. Examples of HVDC transmission systems using voltage source converter technology

Project	Converter type	Rating	Applications	Commissioning year
Estlink (Estonia-Finland)	Two-level	350MW, ±150kV DC	Interconnection and grid re-enforcement	2006
BorWin 1 (Germany)	Two-level	400MW, ±150kV DC	Connection of off-shore wind farm	2009
Cross Sound (USA)	Neutral-Point-Clamped	330MW, ±150kV DC	Interconnection and grid re-enforcement	2002
Murray link (Australia)	Neutral-Point-Clamped	220MW, ±150kV DC	Interconnection and grid re-enforcement	2002
Trans Bay (USA)	Two-switch Modular Multilevel Converter	400MW, 200kV DC	Interconnection and grid re-enforcement	2010
BorWin 2 (Germany)	Two-switch Modular Multilevel Converter	800MW, 300kV DC	Connection of off-shore wind farm	2015

This chapter presents some of the activities carried out within the TWENTIES (2010-2013) research project [20]-[30], co-funded by the European Union, and subsequent developments carried out by partners, addressing issues of DC grid design and control aimed at overcoming some of the technical barriers that still prevent the realisation of complex DC grids.

The chapter starts by presenting the different functions an HVDC grid may fulfil. Subsequently, possible topologies of the future DC grids are discussed, emphasizing the motivations and advantages brought by them. After illustrating converter technology features, affecting key aspects such as protection and interoperability of the DC grid, the operation of a DC grid is introduced. DC protection issues are then discussed. A fully operational DC grid with offshore Wind Farms can be regarded as a large (virtual) power plant capable of providing services to the mainland AC grids: ancillary services will be finally presented.

2. HVDC Grid Functions

Large offshore wind farms should be connected to the onshore power systems via an *offshore grid*. This solution is expected to provide many advantages when compared to point-to-point connections. Moreover, as anticipated in previous section, the long distances and huge power ratings of offshore wind farms call for the adoption of HVDC transmission technology. In fact due to the high capacitance of underground/submarine cable systems, the length of such AC cables for practical use is limited by the charging current of the cable. This drawback can be overcome by using HVDC cables.

The resulting approach is thus an *HVDC grid*. Let us briefly discuss the advantages expected by its development.

1. A major motivation for a grid lies in reliability requirements. These in turn derive from the need to ensure the maximum exploitation of wind energy delivered in electricity market so that the economic investment made for the wind farms brings revenues. Reliability is meant here as the property of being able to assure (at least, partially) the power transfer from offshore wind farms to shore also in case of outage of some components in the offshore grid, thanks to the availability of alternative paths. Reliability requirements concern DC transmission lines and converter stations. For example, in case of point-to-point connection, the loss of one converter station or DC line would imply losing the whole wind farm production. Currently this issue can be partially solved by adopting a bipolar configuration in the converter station.

2.1. Insight

As the capacity of a connected Offshore Wind Farm (OWF) or a cluster of Offshore Wind Farms increases, the limit to the capacity of a VSC based monopole (currently around 1 GW) dictates, first, that either a second monopole or a bipole is required, and then further monopoles or bipoles. Thus, there are at least some degrees of redundancy. In case of a DC grid, there is a diversity of routes which will be less vulnerable to common mode failures (such as anchor dragging) so that the wind energy could still be injected into the grid and arrive at onshore power systems, possibly with some curtailment in case the alternative "route" is congested by other power transactions. The existence of more than one path from an Offshore Wind Farm to mainland grid provides greater availability of energy access in electricity market. In particular, from the short circuit point of view, in a DC grid the fault current can be interrupted isolating only the affected location whereas other sections of the DC grid would remain in service. However, this depends both on being able to locate where the fault is (far from trivial for a DC system) and on activating DC circuit breakers with a sufficient speed. If such localised fault clearance on the DC side is not achieved, the passage of fault current must be interrupted by blocking fault current from the AC side of all terminals of the DC grid. This, in turn, means the loss of all the power being generated on the DC grid or that the DC grid is transferring from one AC system location to another. Once isolation from the AC side has been achieved and the location of the fault on the DC side identified, that location can be isolated by opening one or more disconnectors (that don't need to have fault current breaking capability and so should be relatively cheap, simple and small) and the rest of the DC grid re-energised and, on a DC grid, generation and power transfers restarted. Thus, without DC breakers, in case of faults at particular locations on the DC grid, continuous injection of wind energy on onshore grid and continuous access in electricity market would be impossible: a fault anywhere on the DC grid would require that the whole DC grid is shut down, even if only temporarily.

2. Another driver related to economic and environmental considerations consists of maximisation of wind energy harvesting and market opportunities. In case of radial connection of OWFs, under current market arrangements the only electricity market or market zone to which the OWF can, in the first instance, sell its energy or realise income from a renewables support mechanism is the one operating at the location to which the OWF is connected (the so called 'home' market). If energy is to be sold into a different market, the OWF owner must procure access rights to an interconnector between the 'home' market and the second market. If a price in a remote market zone is to be obtained rather than the local price, financial transmission rights should have been bought. In contrast, if the OWF is connected to a DC grid and two or more of the terminals are connected into different electricity markets or market zones, the controllability of a DC grid allows power to be directed to a chosen AC system location and the maximum power that can be transmitted via any individual terminal is limited by both the rating of the

connection and the capacity of the AC system to which it is connected. Wind power spilling would thus be minimised and best opportunities from the market viewpoint could be exploited.

3. The use of DC grids connecting wind farms far away from each other may smooth out the variability of the wind power, hence achieving steadier power injection into AC grids (mitigation effect) with positive impact on efficiency and security of operation, as well as on the environment. In fact, due to the variable nature of wind, the corresponding power output fluctuates independently from the load. Since the capacity of the off-shore wind farm is large, the magnitude of the fluctuating power is large as well. This might affect the demand-generation balance, leading to the need for procuring large, high-cost power reserve, provided by conventional generators. Some of them could otherwise be kept offline. The mitigation effect may contribute to keep the frequency stability of the AC grid. Of course, statistical studies of wind behaviour are needed in order to make sure that a certain mitigation effect would exist. Such information may support the design and sizing of the DC grid.
4. Security of operation (i.e., continuity of power supply to the users) could also benefit from offshore grids. A wind farm or onshore converter failure results in the loss of power injection into the AC grid. In case of point-to-point connection, stability may be severely affected. In case of DC grid connected to different AC grids, the disturbance may be "allocated" to the individual AC grids, thus reducing the magnitude of the disturbance each AC grid has to cope with. Moreover, the DC grid injections can be dispatched so as to enhance security in case of AC grid faults.
5. In conjunction with wind power production, DC grids could be used to exchange power between the AC grids, thus acting as interconnectors. In this case, some onshore DC terminals could be withdrawing power, while others could be injecting power into the DC system utilizing bi-directional power flow capability of VSC. Power transfer capacity between countries could therefore be enhanced (though the available capacity would be the residual one from wind power, in case of wind priority). Again, this solution would assure greater efficiency of operation and environmental benefits.
6. The interconnection function could also be utilized to provide mutual support between *areas of an AC grid*, or even between *different (i.e., not synchronously connected) AC grids*, for enhanced *security* of

operation with respect to frequency stability and to share *power balancing* resources. In this case the DC grid would act as the means that allows implementation of the support between AC areas, while the offshore wind farms might not be directly involved. In this way, DC systems may partially relieve the requirements for ancillary services (e.g., for some kinds of reserve) to be provided by conventional generators. By this vision, overall operational costs might decrease while at the same time complying with possible future "DC grid code requirements".

7. The above function can be enhanced by making the wind farms contribute to ancillary services, as (downward or even upward) reserve (for frequency stability or balancing) can be shared among the offshore wind generators.
8. In addition to the previous ones, more ancillary services can be provided from the DC grid to the AC grid. The DC grid with wind generation can be seen from the AC grid as a VPP (Virtual Power Plant) able to perform in such a way that the global operation of the integrated DC/AC power system is secure and optimal. Ancillary services are classified and regulated in different ways by the different grid codes and market structures. However, some of them have become common in most grid codes. Conventional control systems of the wind generators and of VSCs are generally required to perform functions such as fault ride through and voltage support during AC network faults; therefore the wind farm grid codes could be adapted to the DC grid. Ancillary services will be discussed in section 7.

3. DC Grid Topology

3.1. Rationale for DC Grid Topologies

Designing a DC off-shore grid on a wide scale requires the development and optimization of economic and technical criteria to be met by this grid in a wide range of operating conditions related e.g., to wind and load level, availability of power plants and grid components, fuel costs, etc. when this DC grid is interconnected to the mainland grid and operated in parallel. Even if an approach for designing a target network is not completely defined yet, it will most likely be built step by step, from existing and planned submarine point-to-point infrastructures (radial connections of clusters of wind farms and submarine interconnectors). Regarding the costs of HVDC devices and the technical complexities to be handled, this is the most likely and pragmatic way to proceed. From a topological point of view, these steps will probably consist in building new capacities and/or extending existing capacity to optimize both wind power transmission and cross-border trade. Therefore it makes sense to define a set of basic topologies that could be, if proven technically feasible, the building blocks for a wide range of possible grid structures in the future.

Offshore HVDC grids are figured in a variety of shapes, depending on the assumptions made and the expected objectives. However, there has been so far limited discussion on how a DC grid should look like, and the rationale for a given layout (or a comparison with other possible solutions). Most DC grid layouts have been driven so far by two alternative motivations:

- minimizing the number of DC components; this is especially true for total cable length or the number of offshore stations, and may stand for a justification of DC grids compared to direct point-to-point connections. In other words, the rationale for some DC grid layouts is only to minimize the investment costs (and, to a certain extent, the exploitation costs for offshore converters). Though this is a decent concern, it should not lead to neglect other aspects such as ancillary services provided to the AC mainland grids, as well as reliability and availability issues
- illustrating some control principles on appropriate DC topologies (that is: layouts that are simple enough, yet suitable to emphasize a specific phenomenon), as the one in the figure below.

This example topology may be adequate to illustrate control principles, yet it seems particularly vulnerable since any contingency occurring on the central tie will prevent the DC grid from transmitting any wind power from a converter to another. This defect is most likely crippling both for grid operators and for wind farms owners; thus, such a DC grid layout seems relevant for academic purpose only.

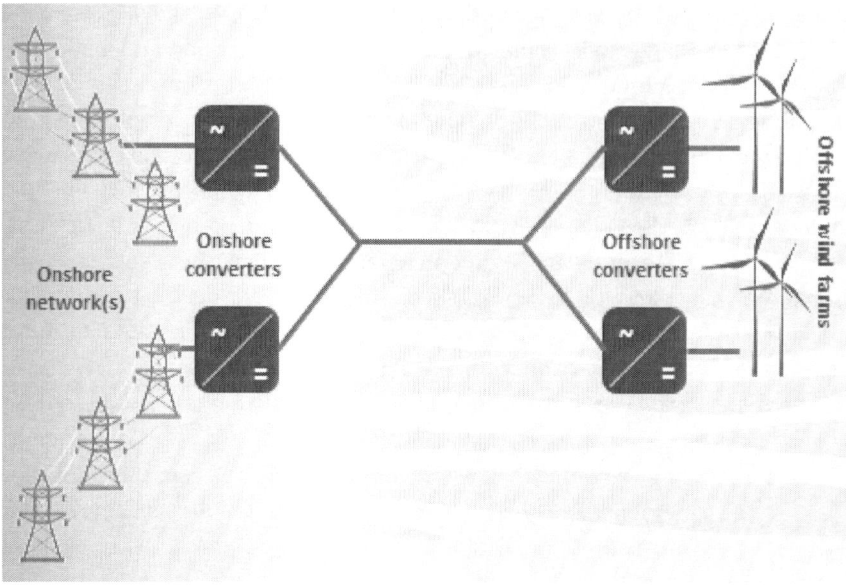

Figure 1. An example of four-terminal DC grid connecting two offshore wind farms (left hand) to two interconnected onshore AC networks (right hand).

As mentioned above, most DC grid studies take for granted (implicitly or not) the fact that offshore grids will be raised *from scratch*, that is regardless of pre-existing offshore wind farm DC connections to the shore. On the contrary, it should be considered that a more realistic philosophy when designing a DC grid is that it will not prevail over existing point-to-point DC connections or be raised as a stand-alone grid: because of the large investments into play, a DC network would be most likely built up step by step, as remarked above, leaning on existing DC connections. Hence, as a starting point, the DC grid that is considered here is built up according to this rationale (as shown in the figure below): assuming the existence of a DC point-to-point connection to transmit offshore wind energy, what would be a reasonable layout to connect another distant wind-farm using a DC connection? In this regard, see the figure below. To address this question, a classification of DC grid topologies is first introduced.

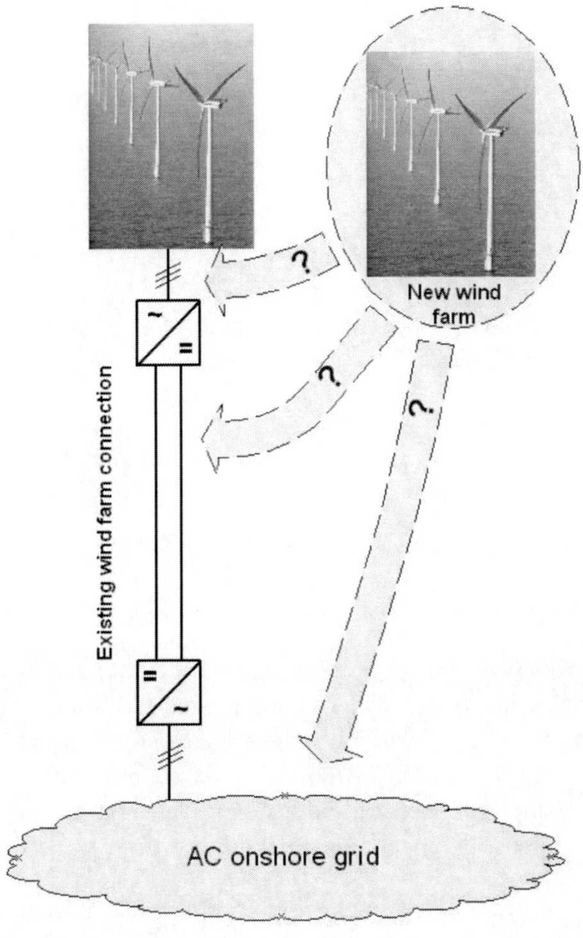

Figure 2. The incremental design of the DC grid: how to connect an upcoming wind farm?

3.1.1. Point-to-Point Connections

Point-to-point connections (see Figure 3) simply connect an offshore wind farm or cluster of offshore wind farms via HVDC to one location on a single onshore AC transmission system. This is the standard way of connecting offshore wind farms and it does not exhibit reliability advantages or flexibility options available with the grids.

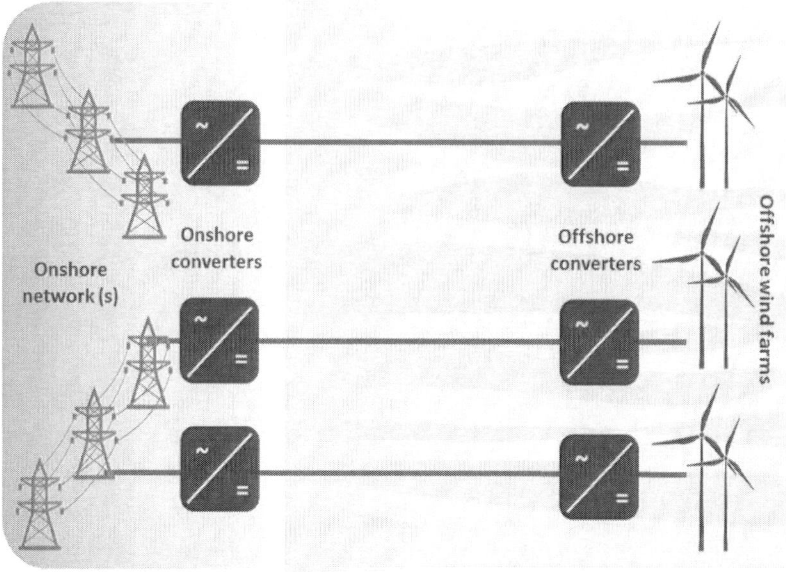

Figure 3. Point-to-point connection of offshore wind farms to shore.

3.1.2. Tree-Like Grids

In tree-like[4] grids, one or more offshore wind farms are connected via a DC grid to one or more AC system onshore location(s), with no possible loop paths within it (see Figure 4); for the sake of simplicity, we refer to them as “backbone” topologies. These grids already present some advantages in terms of increased reliability and operational flexibility with respect to simple point-to-point connections.

When addressing the issue of connecting a new wind farm to form a DC grid, it can be observed that tapping the new one into the existing HVDC scheme is inappropriate due to the risk of cable overload. In fact, the ratings of the initial DC connection were chosen according to the maximal power the first wind farm could deliver. Thus, situations of energy not being supplied might occur.

The most likely first step towards a DC grid is to design another direct point-to-point connection between the new wind farm and the AC network, and to bind both connections with a DC tie which enables power exchange and

[4] «Tree» refers to the term used in Graph Theory (that is: a connected acyclic and undirected graph). Therefore, the tree-like grids are «trees», while meshed DC grids (including meshed backbones described below) are connected undirected graphs with at least one cycle.

mitigation between the two offshore-to-land connections. The resulting "backbone" topology as in Figure 4 implies the realization of intermediate nodes and a cross connection through them.

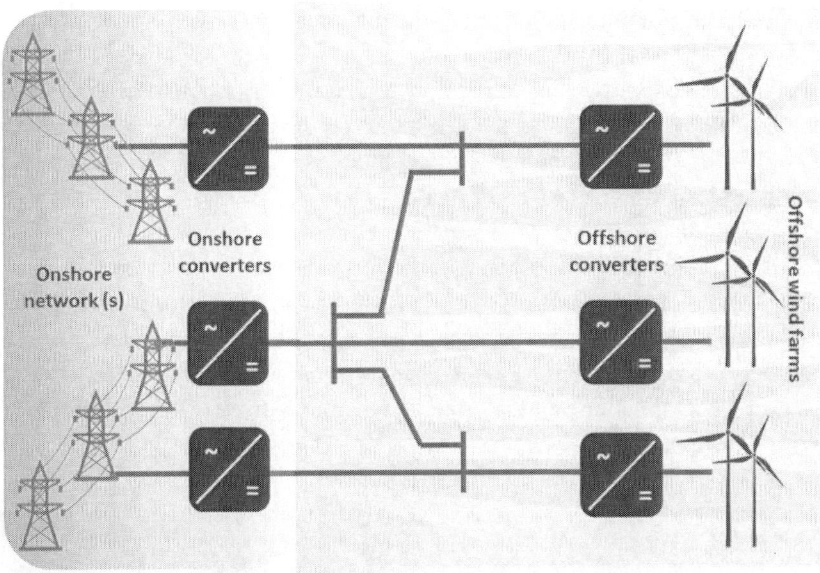

Figure 4. Tree-like grid example (also referred to as the "tree-like backbone").

Assuming the rating of each link is defined to carry only the full power of the one wind farm it is connected to, the loss of a grid side converter would be crippling only in situations where the total wind power from the two offshore ends is superior to the remaining onshore converter capability; however, this would most likely appear during a limited number of hours, according to the duration curve of wind farm generation. In addition, one of the rationales for designing a DC grid is to mitigate the generation from different wind sources, which leads to consider more preferably quite uncorrelated wind generation series, and therefore this reduces the risk of having such a constraint from the generation standpoint.

From a geographical standpoint, this corresponds to the installation of wind farms along a coast shore (similar to the Atlantic Wind Connection project, for instance).

It allows to balance the wind power injections to the grid or to use another path in case a fault occurs on the shore-side lines. The main disadvantage is that if the wind farms are all producing at maximum power, there is no possibility to use the interconnection unless some wind generation is spilt; however, the wider the geographical area, the lower the probability for this situation to occur. Another disadvantage of the backbone topology is that if one DC line connected to the wind farm side VSC is lost, the production of that wind farm cannot be delivered to the mainland AC system.

A simple case of this backbone layout consists of two wind farm connections, hence two onshore converters and two offshore ones; this topology can be referred to as the “H” topology (Figure 5). One variant of this topology is the “star” topology where there is no new cable, but the two existing cables are joined together in a DC node.

In case of more than one onshore connection to an AC system, the connection points may be either within the same AC synchronous area or in different AC synchronous areas. In the former case, the DC grid may be said to be ‘embedded’ within the AC system or to provide inter-area power transfer capability as a complement to the AC network branches. In the latter case, the DC grid can function as an interconnector between the different AC synchronous areas.

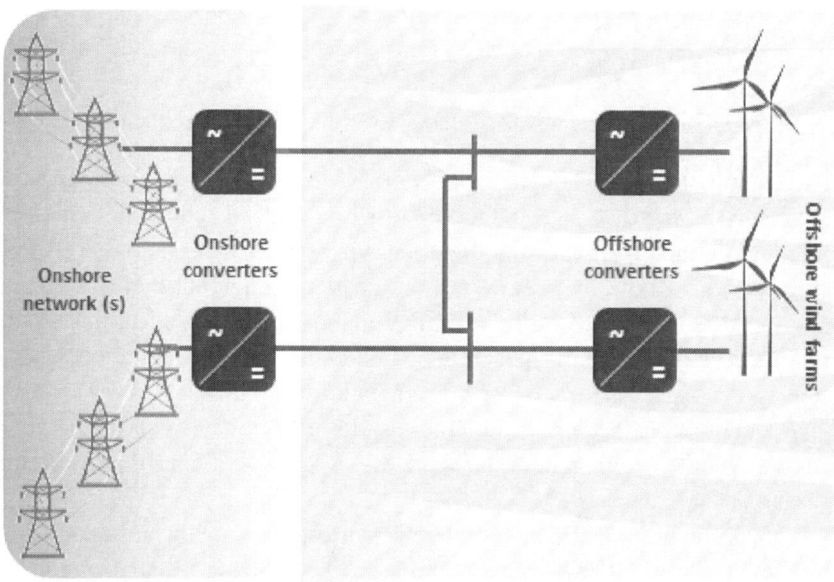

Figure 5. “H” HVDC grid topology.

3.1.3. Meshed Grids

In meshed grids offshore wind farms are connected via a DC grid to one or more AC system onshore location(s), presenting at least one loop path. Reliability of power transfer as well as the capability to provide different services may increase with respect to the backbone case, at the expense of higher operational complexity. Two meshed grids are proposed in particular (see Figure 6): a meshed version of the "backbone" layout, and a mixed topology (including both meshed and radial portions of the grid). Indeed, also the first one has radial and meshed parts; moreover the nodes might be offshore but maybe also at the onshore converter terminals.

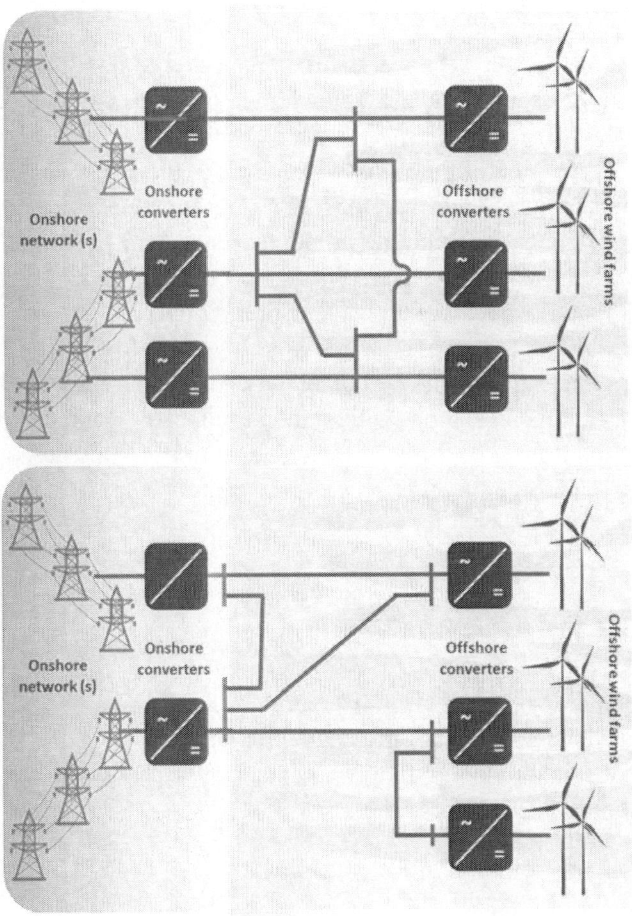

Figure 6. Meshed grid examples.

The meshed grid topology is similar to common AC transmission grid topologies. This topology can be viewed as a target, with many technical challenges still to be met. Due to these many technical aspects still to be handled, it will not be probably built in the near term but this topology has several advantages:

- Energy may use many pathways to go from the wind turbines to the shore, and the same for power exchanges.
- N-1 ([5]) on the DC grid may have a limited impact on the transmission capacity.
- A meshed grid will add flexible interconnection capacity between different countries.
- A large amount of power can be connected to this grid; however, it has to be able to handle a fault without losing the entire generation.

Interconnected wind farms with widely spread locations will benefit from different wind regimes and, therefore, the global output of the grid will be smoothed. The control of multiple points of connection may be designed so that it is possible to manage congestions on both AC and DC grids, using different DC pathflows regarding the situation on the AC grid. This would be the most flexible topology, and can be possibly cost effective by taking into consideration the supplementary reliability and ancillary services it would provide. The cables and stations will be fully used, both by wind power and by interconnection flows. To fully benefit from these advantages, appropriate control regulations for the AC/DC stations and for wind farms have to be designed.

3.2. Criteria for DC Grid Extension

The complexity of the overall controls is greatly compensated by the supplementary benefits the DC grid could provide.

The most suited topology for a grid extension depends on several criteria, which are shortly introduced below:

[5] The expression “N-1” refers to the outage of one component of the power system.

- Impact of topology on possible power flow controls. The power flows, which are derived from Kirchhoff's laws, are impacted by each converter's role (whether in DC voltage control, or in frequency or power control) and their DC connections: DC power flow can be fully or partially controlled, depending on the number of converters used to regulate the DC voltage and the number of DC cables (that is: the meshing of the grid).
- Wind spillage. National or regional grid codes may differ with respect to the capacity requirements to connect a new wind farm. For instance, RTE (the French TSO) has to ensure fully rated connections (i.e., the connection has to be designed for the maximum export capacity of the wind farm). In other countries where wind spillage is allowed, guidelines provide indicators to estimate the optimal connection capacity (lower than the maximum wind farm generation capacity), depending for instance on the wind farm profile. Yet, even if allowed by local regulation or grid codes, wind spillage might prove not to be economically sound based on economic studies intended to estimate the expected losses. The correlation between the wind regimes (which represent the spatial scattering of wind farms) and the grid topology with cable ratings are required as inputs for such studies.
- Interconnection ratings. In contrast to radial DC connections which simply provide offshore wind power retrieval, a DC grid also offers an interconnection capability. Clearly enough, the extension of a DC grid does not question this existing feature, but it may either decrease it (by "saturating" an interconnection segment of the grid to inject offshore power), or increase it by offering alternative or new interconnection paths. As a consequence, the expectations of the extended DC grid with regard to its interconnection capability require thorough investigations of both the grid topology and the ratings of each DC link, to evaluate the expected frequency of capacity saturation and/or the need for wind spillage.
- Impact of N-1 operations. In spite of the limited control they provide in power flow control under normal operation, highly meshed topologies generally provide augmented resilience to N-1 thanks to the number of electric paths remaining in operation. This aspect may be of particular importance in the case of offshore DC grids, where DC cable faults are very likely to be permanent ones.

- Impact on the DC grid stress during a fault. Topology (or more precisely: the meshing and cable length) has a major impact on the stress a DC grid shall withstand during a fault. Hence, a grid extension should be considered in the light of existing and new equipment capability to cope with high and/or fast-rising fault current. In this regard, protection of the DC grid, including DC circuit breaker, plays a fundamental role. The selection for the best-suited topology will also depend on the adequacy between the available capabilities of DC breaker and the most severe expected fault on the parts of the grid that should be protected.

Figure 7 presents a possible evolution from radial links to a meshed DC grid for wind farm harvesting.

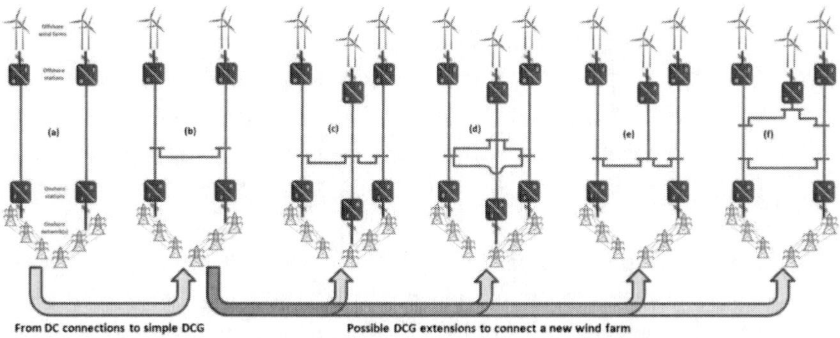

Figure 7. Simple DC connections for two offshore wind farms, and possible extensions to a third one. (a): two point-to-point connections; (b) a simple 'H' grid; two fully-rated extensions: tree-like backbone (c), or meshed backbone (d); two 'H' grid extensions to connect a new wind farm with possible wind spillage: tree-like (e), or meshed (f).

In addition to the non-exhaustive list above, other criteria should obviously be considered for DC grid extension, within the scope of complex cost-benefit analyses. Elements to consider in the analyses are:

- Market integration: What is the additional capacity offered to the market? What would be acceptable for the different stakeholders regarding possible wind spilling or interconnection curtailment?

- Assessment of the reliability of the resulting DC grid (e.g., determine how much power and energy can be delivered in case of unavailability of components)
- Robustness to different events (outage of a cable or of an AC/DC station) in terms of AC/DC system stability, risk of possible adverse interactions in a multi-vendor scheme
- Investment cost of the different solutions (including cables, DC breakers, need for supplementary Power Flow Control (PFC) devices and AC network reinforcement, etc.).

These evaluations require in-depth analyses which are out of the scope of this work.

4. CONVERTERS FOR HVDC GRID

Converter features for DC grids, and the consequences of different converter configurations on DC grid behaviour are dealt with in this section.

Two established converter technologies are currently employed in HVDC transmission systems, i.e., current and voltage source converters.

Current source converters employ thyristors as main switching devices and require an established AC voltage source for synchronization. The two main types of current source converter are the Line Commutated Converter (LCC) and the Capacitor Commutated Converter (CCC). An HVDC system that uses current source converters will be referred to as a conventional HVDC system.

At present, LCC-HVDC transmission systems continue to be used for large power evacuation (>1000MW), where the use of well proven technology supersedes the inherent limitations of this approach.

Generally, dynamic response and control range of conventional HVDC systems based on line commutated or capacitor commutated current source converters are limited by the fundamental switching characteristics of the thyristor devices [31]-[36].

The projected increase in large-scale offshore wind generation and long-distance power transmission has increased the need for high voltage DC networks in which the power can be routed to a number of AC nodes. Such networks present a set of system requirements that challenge the dominance of conventional HVDC technology.

Voltage source converters employ self-commutated devices such as Insulated Gate Bipolar Transistors (IGBTs), with some forms of pulse width or amplitude modulation to control fundamental voltage. These converters allow to independently control the active and reactive power exchange with the grid. Unlike the conventional HVDC system, they can operate stand-alone or in synchronized manner with the AC grid. Voltage Source Converter High-Voltage Direct Current (VSC-HVDC) systems are capable of generating leading and lagging reactive power as synchronous machine when operated in over-excitation and under-excitation modes.

This implies that VSC can contribute to control grid voltage, instead of being a reactive load that needs compensation devices. Moreover, VSC has no short circuit power requirements to operate, i.e., it can "switch on" a de-energised network. VSCs can be built using two-level and multilevel converters.

In recent years, power system industry has recognized that VSC technology is the most suitable converter solution for the development of offshore DC connections.

VSC-HVDC transmission systems are based around self-commutation power semiconductor devices and operate from a regulated DC-bus voltage, which is common to all nodes. This technology allows fast modulation of the HVDC converter terminal AC voltage (phase and magnitude), giving high levels of controllability over the AC voltage and current it presents to the AC grid. VSC-HVDC was first developed for the integration of remote/offshore wind generation. Since then it has evolved rapidly, and its efficiency and power rating have increased significantly; however, these systems still have to match the track record and power capacity of LCC-HVDC [32], [37]-[38].

This section analyses some technical aspects of LCC vs. VSC technology, starting from the requirements associated to the DC grid concept. Features of the two types of converters are reviewed, in order to explain why VSC is more suitable for DC grids.

4.1. Requirements for Converters in HVDC Grids

HVDC converters play a key role in achieving the requirements expected from DC grids. The most relevant features converters have to exhibit are the following [32], [39]-[42]:

1. Ability to provide ancillary services such as:

(a) Fast voltage support, which is critical for operation of weak AC networks such as offshore wind farms, but also essential to aid recovery of the AC grids from AC faults. In addition, this feature has the potential to improve utilization of the AC power network in terms of increased steady-state and transient stability limits, and loss minimization.
(b) Fast power run-back and run-up. This feature could be exploited during AC network faults or loss of major generation or load, to minimize or eliminate active power mismatch (i.e., to provide retarding power) to limit rotor acceleration of the generators. It can also be used to provide fast acting frequency support to AC networks.
(c) Black start capability. This feature could be used to eliminate the need for an offshore start-up generator or STATCOM, hence to reduce costs. It could be used to restore the power to the AC grid following a blackout and, in general, to supply passive AC networks.

2. Power flow control, including:
   (a) Control of power exchange among mainland AC grids.
   (b) DC power management: this may include fast power reversal in some terminals in order to achieve specific objectives such as addressing loop flow and congestion in nearby AC network sections, minimizing total power loss in AC/DC corridors, and redirecting DC power to minimize the loss of power in-feed during a fault in a section of the DC grid.

3. Resilience to AC and DC network faults: this is essential in order to ensure overall system stability during major AC and DC network disturbances.

4.2. Conventional HVDC Systems

4.2.1. Line Commutated Converters

Figure 8(a) shows a schematic diagram of a HVDC station based on the current source Line Commutated Converter (LCC). LCC-HVDC is a well-established and mature technology, with a proven track record of more than 50 years of operation and widely used for large power transmission over long distances. At present the operating voltage and power handling of LCC-HVDC transmission systems reach the impressive level of ±800kV and 6400MW between Xiangjiaba and Shanghai (800kV and 3200MW per pole). The main limitations of LCC-HVDC systems are the following ones [43]:

1. Converter operation requires the presence of an established AC voltage in order to operate (no black-start capability): it is therefore not suitable for wind farm connection [44]. In order to operate as a wind-farm side converter, a LCC converter will require auxiliary systems to establish an active network, such as a self-commutated dynamic reactive power compensation device to provide reactive power support and strengthen the offshore AC network. As an example, the 2×500MW back-to-back HVDC Chandrapur link in India is equipped with 2×426MVAr shunt capacitors in each terminal, configured as four switchable units of 106MVAr for each 500MW converter. As a result, a HVDC converter station based on current source technology is bulky and has a large footprint.
2. LCC-HVDC is susceptible to inverter terminal commutation failure during operation on weak AC networks, and during low real power exchange. Therefore, an LCC converter has a minimum power flow limit, typically 5% to 10% of the system rating. This limit could be extended with the aid of dynamic reactive power compensation devices such as a static VAr compensator (SVC) or a static synchronous shunt compensator (STATCOM) [31][36][45].
3. Converter stations exhibit reactive consumption which varies with the magnitude of the transmitted active power. Variable reactive power compensation is required to aid commutation, and reduce potential over-voltages due to reactive power mismatch. This reactive power compensation takes the form of a large number of high capacity switched capacitors and reactors which are added to the converter station footprint. AC side over-voltages due to load rejection limit the capability of LCC-HVDC system to perform fast power run-back/up. These functional limits may increase the burden on the AC network instead of providing network support.
4. Line commutation of thyristors gives rise to high levels of low order harmonics. These require additional tuned filters to limit harmonic emission into the AC network.
5. Power flow reversal in LCC-HVDC systems requires reversal of the DC link voltage polarity. This has implications for the power flow management in multi-terminal applications, where the change of DC

link voltage polarity is not desirable and cannot be achieved without changing the control mode on all other terminals. Power reversal at an individual converter will require reduction of its current to zero before operation of a polarity reversing switch (this is the so-called hot standby mode). Speed of power reversal is limited by DC cable insulation design, and characteristics of the connected AC networks. Also, it requires sophisticated DC current balancing that ensures that the algebraic sum of all converter DC link currents is equal to zero; this is to prevent unintentional over-voltages on the DC grid. These difficulties may threaten the viability and benefits of the DC grid, and its potential scalability.

6. The use of a large number of switched capacitors increases the risk of parallel resonance with network inductance at non-characteristic low order harmonics.

4.2.2. Capacitor Commutated Converters

The Capacitor Commutated Converter (CCC) shown in Figure 8(b) is an improvement on the LCC, and was developed to extend the operation of the current source based HVDC systems to weak AC networks with a Short Circuit Ratio (SCR)[6] up to 1. It replaces the bulky switched capacitors with fixed tuned AC filters and fixed or thyristor controlled series capacitors [36].
Its major benefits are reduced converter footprint and transformer rating, improved operational stability during AC network disturbances, increased power transfer capability in weak AC networks with a low SCR, and reduced over-voltage at load rejection (this is the sudden reduction of transmitted power due to a fault at the rectifier side, which creates temporary over-voltage at the inverter side due to transient reactive power mismatch, since the inverter reactive power consumption is abruptly changed). Additionally, CCC can change the DC current faster than LCC, without causing significant over-voltages on the inverter AC side as reactive power generation is approximately proportional to the square of the load current flowing through the series capacitors. These features make CCC less susceptible to commutation failure at inverter terminals than LCC. However, both LCC and CCC use mass-impregnated cables which are heavier than extruded cables used with voltage source converters (fast and easy to join and install). Overall CCC suffers most

---

[6] The SCR is defined as the ratio of the short circuit power (MVA) of the AC system at the AC bus to the rated DC power at that bus.

of the limitations of LCC, although to a lesser extent. Therefore, its adoption in the DC grid with a high penetration of wind power may compromise most of the DC grid potential benefits previously presented. However, its use for a DC grid with all the terminals connected to strong AC networks, with power flow in one direction such as that currently under construction in India (±800kV North-East Agra link, with 8,000 MW converters), could be justified (the priority is bulky power transfer over long distances).

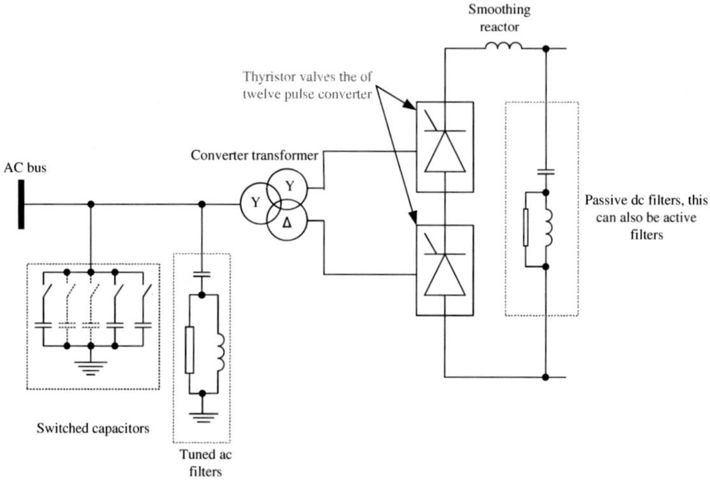

(a) Line commutated converter

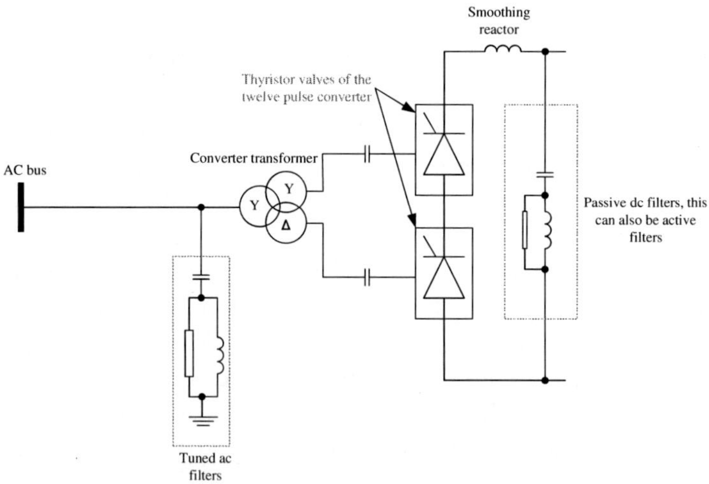

(b) Capacitor commutated converter

Figure 8. Schematic diagrams of conventional HVDC stations based on natural and capacitor commutated converters.

4.3. VSC Based HVDC Systems

The introduction of pulse width modulated Voltage Source Converter technology into HVDC transmission systems has increased their viability in many applications in terms of cost and performance. The main benefits of VSC-HVDC over classic LCC-HVDC are [32],[41],[46],[47]:

- Converter inherent leading and lagging reactive power generation capability (resulting in smaller converter size and reduced filtering requirements).
- Independent control of active and reactive power; this allows VSC-HVDC links to provide damping, and fast frequency and voltage support to AC networks without compromising system performance.
- Black start capability, which extends the use of HVDC systems to weak AC networks with no generation.
- Achievement of fast power reversal with no need for DC link voltage polarity reversal (zero active power exchange can be achieved, without additional devices as with LCCs and CCCs). This enables to use inexpensive cables and transformers with lower insulation requirements, as these are not required to withstand high voltage stresses during power reversal.
- Improved AC fault ride-through capability: this may enable the HVDC terminals to support the AC networks during faults, hence enhancing transient stability of the overall power system.

VSC-HVDC transmission systems can be built using two-level or multilevel converters. The two-level converter approach suffers from high semiconductor losses (mainly, switching losses), and exposes the interfacing transformer to high voltage increase, with high *dv/dt*, due to switching of large voltage steps at high frequencies between 1kHz and 2kHz. Harmonic Filter requirements of two-level VSC are less than those of LCC but remain significant (ranging from 20% to 33% of the system rating).

In the multilevel approach, the neutral-point-clamped converter (NPC-VSC) or modular converter could be used. However, HVDC systems based on the NPC-VSC do not maximize the benefits of multilevel modulation such as the low *dv/dt* and the low effective switching frequency per device. Also they are vulnerable to asymmetrical AC faults, such as open-circuit faults, and single pole-to-ground DC side faults. For these reasons, the practical application of NPC-VSC-based HVDC transmission is limited.

The introduction of modular multilevel converters into HVDC transmission systems have brought several advantages, such as low conversion losses (high efficiency); low switching frequency per device (low audible noise, therefore the converter station could be placed outdoor); the production of high-quality AC voltage with low *dv/dt* and harmonic content (this reduces voltage stresses on the converter transformer and eliminates the need for phase interfacing reactors and AC side filters); and resilience to AC and DC faults. Additionally, modular multilevel converters do not use input DC link capacitors as two-level and NPC converters do; they use instead a number of distributed DC capacitors to create multilevel voltage waveforms. This elimination of DC link capacitors has benefits for DC circuit breaker design as the let-through current and current breaking capacity are greatly reduced. These features make VSC-HVDC systems more attractive for multi-terminal DC grid than LCC or CCC-HVDC systems are.

4.4. Review of VSC Converters for HVDC Grids

This section presents a brief review and high-level discussion on established and emerging voltage source converters for high-voltage DC transmission systems.

Figure 9 summarises most of the existing VSC topologies. As DC grids are expected to become a reality, HVDC transmission systems experts have recognized that the converter topologies that adopt lumped DC link capacitors such as in Figure 9 (a), (b) and (e) increase the DC fault current transient component due to discharge of the DC side capacitors (reservoir capacitor and DC cable distributed capacitors). Therefore, DC circuit breakers must be able to withstand high let-through currents, have a high current breaking capacity and be sufficiently fast acting.

In an attempt to address such shortcomings, most present and future HVDC manufacturers have adopted or are developing new generations of chain link converter topologies that can operate without DC link capacitors, such as half and full bridge modular multilevel converters. To further improve the system transient response during or following DC side faults, reverse blocking converters such as full-bridge modular, alternative arm modular, and hybrid cascaded converters can be used to reduce the current stresses in converter semiconductor switches by eliminating grid contribution to DC fault.

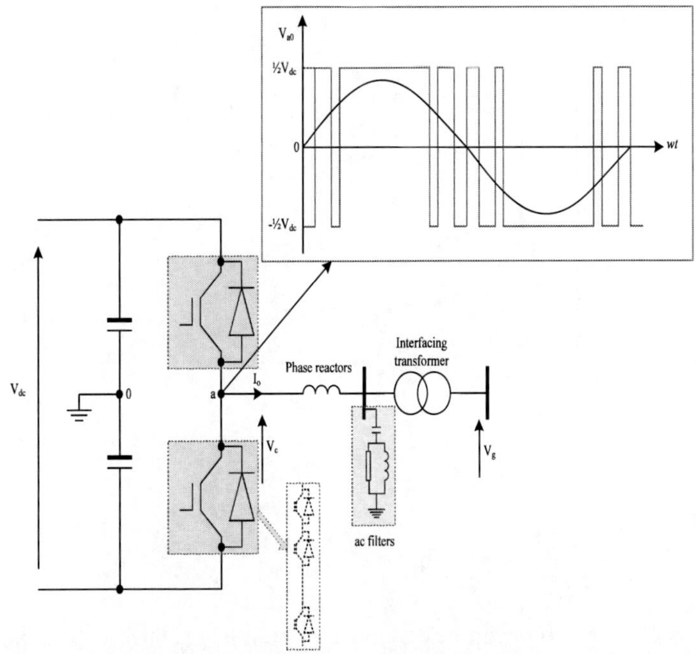

(a) One phase of a two-level converter.

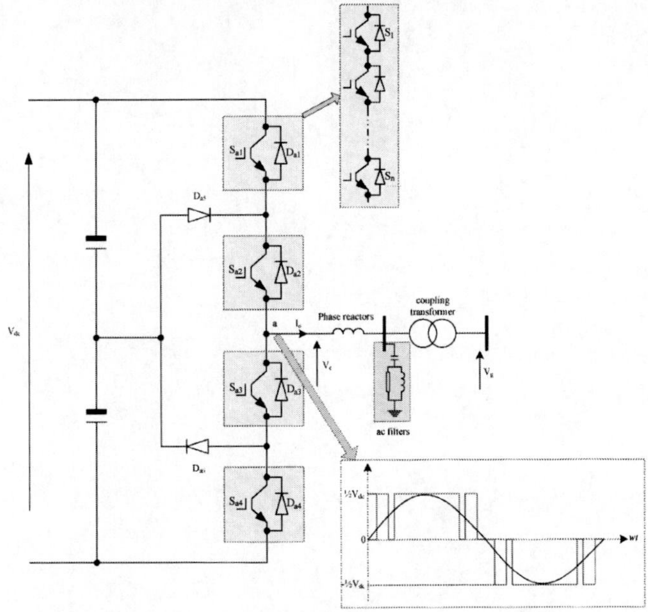

(b) One phase of a neutral-point-clamped converter.

Figure 9. (Continued).

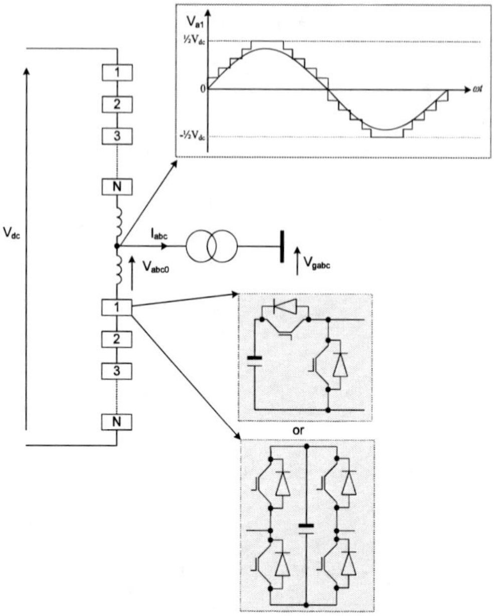

(c) One phase of a half or full-bridge modular multilevel converter also known as cascaded two-level converter.

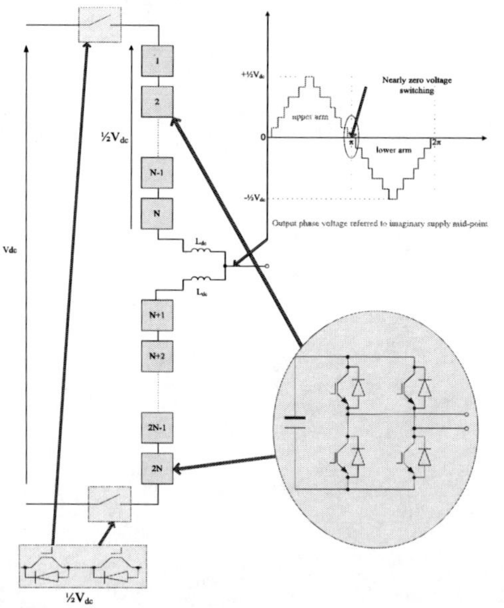

(d) One phase of an alternative arm modular multilevel converter.

Figure 9. (Continued).

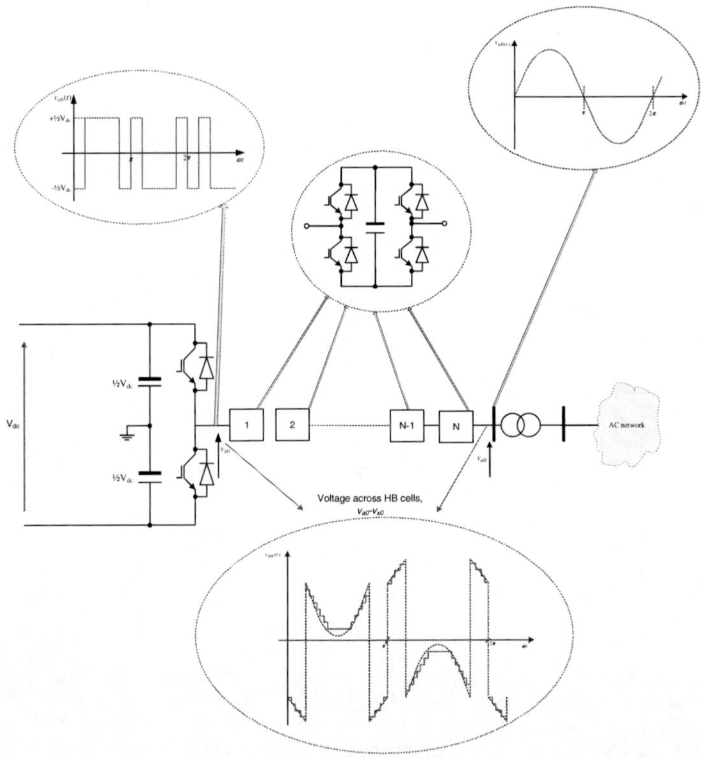

(e) Hybrid cascaded multilevel converter with AC side cascaded full-bridge cells

Figure 9. Main conventional and emerging VSCs for high-voltage DC transmission systems.

By stopping grid contribution to the DC fault current, the steady-state DC fault current is reduced to zero. However, most of the present manufacturers are reluctant to adopt this type of converters due to high conversion losses.

Table 3 summarises the relevant features of the converters in Figure 9.

The sequel of the subsection describes representative voltage source converter topologies for HVDC transmission systems already in operation such as the two-level, the neutral-point-clamped and the modular converter, and those currently under consideration.

Table 3. Global comparison among converter types

	Two-level converter	**Neutral-point-clamped** converter	Half-bridge modular converter	Full-bridge modular converter
Reactive power capability	Limited only by available DC link voltage	Limited by available DC link plus DC offset for capacitor voltage balancing	Limited only by available DC link voltage plus arm inductances	Limited only by available DC link voltage plus arm inductances
Bidirectional power flow	100% in both directions	100% in both directions achievable, but subject to modulation index margin for capacitor voltage balancing	100% in both directions	100% in both directions
AC filters and interfacing reactors	Relatively large AC filters	Relatively large AC filters	AC filter and interfacing reactors could be eliminated	AC filter and interfacing reactors could be eliminated
Conversion losses	Very high (mostly switching losses)	High (mostly switching losses)	Low (mostly on-state losses)	Moderate (mostly on-state losses)
AC fault ride-through capability	Excellent	Excellent, but, in bi-polar configuration case with DC link neutral-point grounded, AC side filters and interfacing transformer at neutral-point in converter side must be grounded.	Excellent	Excellent
DC fault ride-through capability	Poor	Poor	Good	Better

4.4.1. Two-Level Voltage Source Converter

Figure 9 (a) shows a one-phase leg of a two-level converter. It is capable of generating output voltage with two voltage levels $\frac{1}{2}V_{DC}$ and $-\frac{1}{2}V_{DC}$ between 'a' and a virtual mid-point ground '0' as shown in the figure [44]. The two-level converter can be controlled using the main stream modulation strategies such as carrier pulse width modulation, selective harmonic elimination modulation and space vector modulation. The first two modulation strategies have been used intensively in the HVDC transmission systems due to their robust dynamic performance during network disturbances. As the maximum operating voltage of a single IGBT is limited to 6.5kV at present, the two-level converter HVDC systems use series connected IGBTs in each arm to enable operation at high voltage levels such as ±150kV and ±300kV. Power quality considerations require device switching frequencies in the order of 1-2kHz. This results in high switching losses and exposes converter transformers to high voltage stresses from fast switching between the positive and negative HVDC voltage. Therefore, two-level VSC systems require a converter transformer with high insulation requirements to withstand high voltage stresses, and relatively large AC filters at the converter output. A DC chopper in the DC side is normally used to improve system AC fault ride-through capability against loss of grid and AC side faults in close proximity to the converter bus. In order to reduce the risk of device failure during the DC faults as a result of the AC grid contribution to the DC fault current through freewheeling diodes of the IGBTs, the reactor between the Point of Common Coupling (PCC) and the converter terminals must be sized large to limit the magnitude of this current. This may reduce converter reactive power capability as a result of reduction in the available DC link voltage.

4.4.2. Neutral-Point-Clamped (NPC) Converter

Figure 9 (b) shows the circuit for a one-phase leg of a Neutral-Point-Clamped (NPC) converter. This circuit generates three voltage levels $\frac{1}{2}V_{DC}$, 0 and $-\frac{1}{2}V_{DC}$ per phase between 'a' and neutral point '0' as shown in the figure. Proper operation of this circuit requires the voltage drop across the upper and lower DC link capacitors must be maintained at $\frac{1}{2}V_{DC}$. The voltage drop across each individual device is limited to one capacitor voltage [48]-[51]. Neutral-point-clamped converters suffer from poor utilization of semi-conductor devices and non-distribution of semiconductor losses. Loss distribution and semiconductor utilization tend to improve at high modulation indices (0.8~0.9), where most of the HVDC transmission systems are designed to operate, and deteriorate at low modulation indices. Active Neutral-Point-Clamped (ANPC) converters have been developed to address this issue, where each clamping diode is replaced with a composite device which includes IGBTs and a freewheeling diode. The NPC and ANPC converters generate output voltage with lower harmonic content and *dv/dt* and produce lower semiconductor losses than the two-level converter does. Therefore, filtering and converter transformer insulation requirements may be reduced compared to those for two-level converters. HVDC systems which use NPC converter technology can be controlled using carrier based pulse width modulation, selective harmonic elimination (or minimization) and space vector modulation. Space vector modulation, in particular, is suitable for control of NPC and ANPC converters [52][53]: in fact, the DC link capacitor voltage balancing of these converters is normally influenced by the current contribution from the three phases, and space vector modulation uses these currents in the redundancy function to maintain the voltage balance of the DC link capacitors. However, in applications such as HVDC transmission systems where continuous operation during unbalanced conditions and different AC faults is required, the space vector modulation has practical difficulties to meet such requirements. Therefore, the carrier based pulse width modulation and selective harmonic minimization are preferred methods for controlling NPC converters. As the neutral-point-clamped converter uses the DC link capacitors as an energy tank, the DC link capacitors of the NPC converter must be sized large enough to reduce the variability of the devices voltage stress, which can be caused by the neutral-point voltage fluctuation. This may influence converter station footprint. The resiliency of HVDC systems based on NPC converter against loss of grid or AC faults can be improved using a DC chopper as in the two-level converter case. The DC fault ride-through capability of the NPC converter can be improved in a similar manner as in the two-level converter case by increasing the effective impedance between the PCC and the converter terminals. The other disadvantage of the NPC converter is that 100% active power transmission in both directions may not be granted, because of potential conflict between the needs of the AC voltage controller (which increases the modulation index in order to produce more reactive power) and the need for capacitor voltage balancing (which requires modulation index margin to generate an appropriate DC offset for balancing purposes).

4.4.3. Modular Multilevel Converter (M2C)

Figure 9 (c) shows a one-phase leg of an m-level modular converter, where m=N+1 and N is the number of cells per arm. It can generate *m* voltage levels per phase as shown in Figure 9 (c). In this converter, voltage across the cell capacitors of each arm must sum to the total DC link voltage minus a small voltage drop in the arm inductance at any instant [54][55]. The voltage across each cell capacitors is $\frac{1}{N}V_{dc}$ and each switching device experiences maximum voltage stress limited to one capacitor voltage. The arm inductance L_{DC} is normally sized to limit potential inrush current during the capacitor voltage balancing process and circulating currents between the phases during unbalanced operation. Also it acts as a low pass filter contributing to the improvement of the phase current quality. Larger arm inductances may introduce distortion in the output voltage. Also it may reduce the available DC link voltage for reactive power compensation. Modular multilevel converters (acronym MMC or M2C) have the combined advantages of the flying capacitor and cascaded multilevel converters such as phase voltage redundancy, modularity, and resiliency to AC faults. Also it has a set of new features such as [54]:

- Extendibility to a high number of levels without capacitor voltage balancing problems.
- No reservoir capacitor: this may allow relatively fast recovery from DC side faults compared to two-level and NPC converters because the risk of device failure from over-current during DC link voltage build up following system re-energization is significantly reduced. This may also permit the use of a DC circuit breaker with lower current breaking capability.
- Ability to operate over all power factors (0~1) and modulation index linear range (0~1.15).

These features make modular multilevel converters well suited to high-voltage application and superior over two-level, neutral-point-clamped and flying capacitor converters in terms of quality of output voltage (no AC filter required), low switching losses, low dv/dt (it is possible to use cheap transformers with standard insulation, such as normal power system transformers in transmission and distribution systems) and low audible noise as the converter switching devices operate at extremely low switching frequency such as 150Hz or less. Notwithstanding all these advantages, large numbers of cell capacitors and arm inductances may increase converter footprint, and the existence of the DC component in the converter arms currents and flowing through the semiconductor devices raises some concerns about the level of semiconductor losses in high-voltage high-power applications such as HVDC where the losses are expensive. Because of the phase voltage redundancy feature, the carrier based pulse width modulation and amplitude modulation that operate in a similar way to fundamental switching modulation but without the need to calculate any switching angles are well suited for modular multilevel converters with a large number of levels. The DC link utilization by the modular converter can be improved using standard techniques available in the literature, such as third harmonic injection (THI), switching frequency optimal (SFO) modulation and selective harmonic minimization.

The features of the modular multilevel converters have considerably attracted the main manufacturers of HVDC transmission systems, who have started to develop new voltage source converters HVDC systems using this technology.

4.4.4. Full-Bridge Modular Multilevel Converter (FB-M2C)

Figure 9 (c) shows a one-phase leg of a full-bridge modular multilevel converter with N cells per arm. The operation of this converter is similar to that of the two-switch cells modular converter previously discussed, except for the use of the full-bridge cells which introduces new sets of redundant switch states in the phase voltage: these states allow the voltage error in the cell capacitors to be corrected within sub-cycles of the fundamental period [56]. This permits the use of small cell capacitors, hence a small footprint. The use of full-bridge cells in the converter arms increases semi-conductor conduction losses but achieves the important feature of reverse blocking capability. This feature allows the converter to block the current flow in the converter switches (IGBTs and their freewheeling diodes) during the DC faults, hence there is no active and reactive power exchange with the AC system (zero current, zero active and reactive power exchange).

As a result, the risk of device failure due to increased current stresses during DC faults is significantly reduced. Also this may eliminate the need for fast DC circuit breakers and may allow the use of DC circuit breakers currently used with conventional HVDC systems to isolate the DC fault. Such an improvement in the converter design may contribute to spread VSC-HVDC transmission system technology with overhead lines (subject to transient faults due to lightning) and accelerate the practical realization of multi-terminal DC systems based on voltage source converter technology.

The converter in Figure 9 (d) is proposed as an alternative to HB-M2C and developed with the main objective of achieving DC fault reverse blocking capability at reduced semiconductor losses [57]. This converter is called alternative arm M2C. The number of H-bridge cells required per arm is half of that in the HB-M2C. Each converter arm operates for 180° only. The director switches in Figure 9 (d) are used to increase the voltage blocking capability of each arm and to facilitate zero voltage switching during current commutation from one arm to another (for further reduction in switching losses).

4.4.5. Hybrid Multilevel Converter with Series Full-Bridge Cells in the AC Side

Figure 9 (e) shows a one-phase leg of a fault tolerant voltage source converter with series full-bridge cells in the AC side. The main two-level converter (high-voltage stage) uses series connected gate commutated switching devices such as IGBTs, capable of blocking the full line-to-line DC voltage. The series full-bridge cells use low voltage rated IGBTs with a maximum voltage stress limited to the cell capacitor voltage. The proper operation of the proposed circuit requires robust capacitor voltage balancing to ensure that the voltage across each cell capacitor does not exceed the switch device rating under any conditions. The converter in Figure 9 (e) uses the two-level converter to synthesize the desired fundamental voltage component by switching between $+\frac{1}{2}V_{DC}$ and $-\frac{1}{2}V_{DC}$ according to a predefined pattern. The full-bridge cells act as a series active power filter to attenuate the low order harmonics in the output phase voltage and create a multilevel voltage waveform with voltage steps equal to the cell capacitor voltage, in order to reduce the voltage stress on the winding of the interfacing transformer (low *dv/dt*). The distinct feature of the circuit in Figure 9 (e) is that the switches of the two-level converter and H-bridge cells are operated with relatively low switching frequency; this results in low switching losses.

4.5. Concluding Remarks

HVDC networks could be built around LCC and CCC converters as such an approach has a number of attractive features, including use of established technology, low losses and resilience to DC side faults. However, these converters impose significant technical and operational restrictions, which could limit the functionality of the DC network. In contrast, VSC based HVDC is well suited to multi-terminal operation, as VSC stations can operate from a common regulated DC bus voltage with fast acting power control at all terminals. In particular:

- Line and capacitor commutating converters are unsuited for multi-terminal HVDC networks and connection of weak AC networks such as off-shore wind farms, for the following reasons:
  - ➢ Power reversal requires complete system reconfiguration (power circuit and control systems).

  - The burden on the AC network is increased where the inverter terminal is connected, and this increases the risk of commutation failure.
  - As the number of terminals increases, control complexity becomes prohibitive, especially in terms of current balancing and recovery strategies from AC and DC network faults.

- Voltage source converters are well suited to multi-terminal HVDC networks. There are several voltage source converter topologies presently operational in real world HVDC transmission systems, namely two-level and (NPC or ANPC) converters with series device connection, and modular multilevel converters (M2C). The main attributes of these topologies can be summarized as follows:

  - Two-level converter based series device connection is well established, but its main drawbacks are high semiconductor losses, need for large AC filters, need for interfacing transformer with high insulation to withstand high *dv/dt*, and vulnerability to DC side faults.
  - Half-bridge modular multilevel converter reduces semiconductor losses and does not require AC filters, but it remains vulnerable to DC side faults, and the use of individual cell capacitance may lead to increased converter footprint.
  - Although H-Bridge M2C suffers from increased semiconductor losses, it retains most of the attributes of the half-bridge M2C, and it offers DC fault reverse blocking capability.
  - Other topologies are being investigated which provide different levels of performance optimization; however, these are still at the development stage.

A final important note concerns interoperability, i.e., the possibility to operate, in the same DC grid, HVDC converters realised with different configurations, possibly by different vendors. Although the internal operation of the different converter types may be radically different, their outer control loops can be configured to provide the same system level functionality. Studies performed in the TWENTIES project indicated possible interoperability of converters under steady state, dynamic, and under AC fault conditions. Although preliminary studies indicate no fundamental limits on compatibility, it should be noted that the design details of each converter will dictate individual dynamic responses. Where networks are built using a range of technologies it will be necessary for network operators to provide standardised requirements for system level converter behaviour. More significant interoperability issues may be encountered when DC faults are considered. Two-level inverters will give rise to more severe DC fault transients and may incur longer re-energization times than M2C converters. Reverse blocking converters can actively manage the AC input to DC faults; however, this may require the same functionality at all converter stations.

5. Operation of the Offshore HVDC Grid with Wind Farms

5.1. Overview

Similarly to AC grids, DC grids must be equipped with control, protection, and defence systems, aimed to operate the grid under steady state and disturbed conditions.

Because of the ability of HVDC converters to control active power, the DC grid presents some degrees of freedom in dispatching the offshore wind generation onto the onshore injection terminals. Therefore, optimal allocation of injection dispatch to onshore grid can be pursued according to market rules. In fact, the initial dispatch of the DC grid injections into the AC bulk power system depends on the offshore wind generation profile, combined with considerations of day-ahead market prices and security constraints on the overall AC/DC system. The DC grid could be connected to different market areas, hence it may be convenient to inject as much power as possible into the market area exhibiting the highest price. In this regard, it is worth mentioning that the tools currently used by power market operators should be adapted in order to account for this enhanced flexibility.

Steady-state controls provide the transfer and interconnection capabilities of a DC grid, to allow the desired power dispatch among DC grid terminals. "Pure" steady state conditions are an abstraction, because small changes continuously occur in the power system. In the DC grid for offshore wind farm connection, small changes are primarily associated with wind power fluctuations. We can refer to these conditions as quasi-steady state. DC grid control must be able to accommodate the variations in order to keep operational quantities within allowed technical limits, with the aim of guaranteeing security of operation. These small variations must be managed by the DC grid control in a fully automatic way.

Besides random fluctuations around an average value, system quantities can exhibit slow, but definite trends of change. In the DC grid, this situation occurs when the wind is ramping up or down. In this case, the automatic controls aimed to adjust small variations may lead to operating conditions of the DC grid or of the AC grids, which are not adequate from the security standpoint, or are not consistent with market arrangements (e.g., relevant to balancing). Thus, the control room operators have to implement control actions, if needed, aimed to preserve adequate security margins possibly according to market outcomes. As far as actions on the DC grid are concerned, they will typically consist in redispatching the power flows in the DC grid by acting on the onshore converter power injections and, only if strictly needed, in reducing the offshore wind power generation. Indeed, HVDC grid operation introduces new requirements for control room analysis tools and poses complex coordination issues among TSOs. Current tools for security assessment, both offline and online, need to be adapted to account for the DC grid interacting with one or several AC grids. Moreover, control solutions have to be agreed between the different involved parties on a wide scale and involving suitable market and regulatory frameworks.

Sudden, large events, such as short circuits or equipment outages, may also occur. These events trigger severe disturbances and complex dynamic evolution in the system, which may lead to power infrastructure damage or operational security issues unless suitable countermeasures have been previously designed. In particular, faults need to be cleared within very short time scales (few milliseconds) by isolating the faulted component. This task in charge of protection systems is more challenging than in AC systems due to the shorter allowable time and much higher difficulty of interrupting DC fault currents. On the other hand, selectivity is a mandatory requirement in a DC grid, i.e., only the minimal set of components must be disconnected in case of fault, in order to assure continuity of operation of the DC grid itself as much as possible. "Switching off" the DC grid would potentially imply catastrophic disturbances on the AC grids in case of high power levels (e.g., several GW's) are involved, due to the resulting, severe and sudden unbalance.

Selectivity and the other properties required to protection systems (such as dependability, security) are not enough to assure security of operation of the overall AC/DC grid[7]. In fact, the AC/DC grid must be operated in such a way that any foreseen component loss or fault, cleared by protections, does not raise security issues. This would be the case, for instance, if the loss of power injection from an onshore converter caused instability or an overload in an AC or DC branch. In order to prevent security problems, the system must be operated in the pre-contingency situation with suitable security margins.

In order to enhance the strength of the system with respect to AC and DC contingencies, thus allowing the system to be operated in more stressed pre-contingency conditions, special controls can be designed. Some of them can be classified within the ancillary services, such as the Fault Ride Through (FRT) capability. Other controls are closer to the so-called defence schemes, broadly defined as automatic systems aimed to prevent an incipient instability by triggering actions, such as load shedding, according to a defined logic.

The present section introduces steady state controls of the DC grid with offshore wind farms. Afterwards, some concepts related to the dynamics of the DC grid are provided. Finally, security issues are dealt with. These concepts will also be useful to understand protection and ancillary services, which are presented in the next sections.

5.2. Accounting for Wind Fluctuations in DC Grid Control

As recalled above, the role of the DC grid with respect to power transmission is twofold: it shall transmit offshore power to the AC network; additionally, it may also play the role of an interconnection between different AC onshore zones (for planned power exchanges, for instance). The conjunction of pre-defined and variable power injections over time requires appropriate coordination of power exchanged at each terminal of the DC grid, even under normal operation.

To this purpose, voltage droop control is proposed as a general principle used to satisfy the overall dispatch energy balance in a DC grid (though this is also widely applied for other power system equipment). Unlike conventional schemes, the power flow across the DC link is not precisely controlled by a Dispatch Center via a setpoint: each converter running the droop control has

[7] In the power system community, the term "security" refers to the ability of the system to withstand a specified set of disturbances, without jeopardising the continuity of customer supply. The most popular security criterion is named "N-1", meaning that the system must be robust with respect to any single component loss.

autonomy, so that the load flow changes to balance the generation and demand variations. Indeed, in a DC grid scheme, each converter is allowed to independently adjust its terminal DC voltage following a characteristic, which generally consists in a DC voltage/DC current, or in a DC voltage/DC power slope.

The existing point-to-point schemes connecting offshore wind power to shore rely on simple voltage/power controls which enable basic coordination between the two ends of the DC circuit. Under this scheme, one converter controls the DC voltage, and the other controls the DC power from the wind farm. This control simply ensures power balance in the DC systems, whatever the wind speed.

In the case of a DC grid, the straightforward extension of the previous strategy would consist of having one terminal performing voltage control, and the other onshore converters performing power control. However, this solution does not seem adequate for reliability and control performance reasons: offshore wind power variations would be absorbed by a single onshore converter acting as "slack" node; moreover, network stability issues could occur in case of loss of the unique voltage controlling VSC. Therefore, different control strategies are envisaged: DC Voltage-Power droop control, to be applied to several onshore DC/AC converters acting as "distributed slack" for active power, appears a good solution. The offshore converters, instead, would control the voltage and frequency of the AC island, thus injecting into the DC grid all the available power from the wind farms. This droop principle also ensures the overall balance of power injections during normal operation *for any DC grid topology*.

Three different types of nodes can thus be identified in a DC grid: wind generation nodes, transit nodes, HVDC "grid side nodes" (namely, at the interface with the mainland AC network). Based on the above discussion, three alternative control types can typically be assigned to the DC nodes:

- V-controlled node: voltage is set to a setpoint V_{ref}. It acts as the slack bus of the DC network, in case of "lumped" slack;
- P-controlled node: power injection P is set as an input. Typically wind generation nodes belong to this set. Transit nodes are not inverter-controlled, but they obviously belong to the same set;
- Voltage-Power (VP) droop node: voltage V and power P at the node are controlled via the following strategy: $V = V_{ref} + k\ (P - P_{ref})$. This control strategy will typically be applied at the HVDC interface nodes.

Moreover, a node may undergo current injection limitation: if the current I through a VP droop controlled converter exceeded its maximum allowable current, the converter would shift to a current limitation control mode. The new relationship between voltage and power is: $V = |P| / I_{MAX}$.

Figure 10 reports the VP droop characteristics with current limitation.

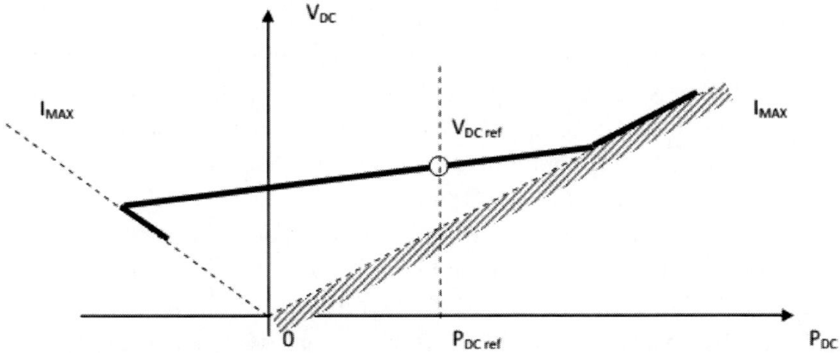

Figure 10. PV droop control in HVDC converters.

It is worth pointing out that telecommunication is only required to ensure optimal operation of the DC grid by adjusting control setpoints from time to time. Therefore, there is no need for fast communication as far as coordination is concerned; furthermore, the voltage droop principle guarantees autonomous safe operation in the case of communication loss, at least as far as the wind power injections do not vary “too much”.

5.3. Dynamic Phenomena in the DC Grid

In AC systems, the control of voltage and power is essentially performed by synchronous machine and prime mover controls of conventional generators. Power flows in the lines are generally not directly controlled (Ohm’s law applies to line parameters) as they depend on the generators’ power dispatch. Phase Shifting Transformers (PST) and some FACTS devices can control active power flows. Though the number of PST installations is steadily increasing, power flow control devices are not widely deployed.

The DC grid exhibits different characteristics, in particular faster time constants and high controllability at converter injection nodes. Converters may control voltage and/or active power at the DC side, as recalled above.

The DC grid presents complex interactions with the different AC systems it is connected to, namely wind farm islands and mainland grid. In designing DC grid control and protection systems, coordination is fundamental in order to ensure overall system stability. Indeed, disturbances may propagate from the mainland AC system to the DC grid and the offshore wind generators, and vice versa. It is therefore important to operate the system in a secure way, considering the mutual interactions between AC and DC.

A basic principle in power system operation is the "instantaneous" balance between power production and consumption (loads and losses). If this balance is not met within strict margins, instability phenomena will occur, leading to out of range electrical quantities and eventually to blackout.

Disturbances like the loss of onshore VSC converters or AC faults on mainland grid may cause severe active power imbalances between the sending end converters and the receiving end converters of DC grid, raising security issues. In a DC grid, power imbalances are reflected by the voltage magnitude instead of the voltage phase angle as it happens in AC systems. For example, when the AC grid experiences an AC fault near a Grid Side VSC (GS-VSC) i.e., the converter that is located onshore, the voltage of the GS-VSC's Point of Common Coupling (PCC) drops. The power delivered to the AC grid will also drop if the converter current controller reached its limit. Consequently, the GS-VSC will partially lose its ability to control the DC voltage. The excess of power produced by the wind farm is absorbed by the capacitances of the DC grid causing overvoltages within few milliseconds in the DC grid itself. To relieve the overvoltage, Wind Farm side VSCs (WF-VSCs) will change into a DC voltage control mode, which will rapidly limit the power injection into the DC grid. The consequent abnormal AC voltage in the wind farm islands, due to the change of control mode on WF-VSCs, might cause the tripping of the wind turbines due to low voltage protection. Thus, WF-VSC control during AC grid faulted condition should be coordinated with the low voltage fault ride through (FRT) capability of the wind turbines. Also, the reduction of AC voltage of the wind turbines will result in generator over-speeding. In this case, to balance the power of the wind turbines, a load resistance, such as a dump resistor[8], may be switched on to rapidly dissipate the extra energy and the slower pitch angle regulation can subsequently be used to reduce the captured wind power.

[8] Sometimes also referred to as crowbar.

The sudden loss of an onshore power converter leads to a fast decrease of the power injection capability of the DC grid to the mainland AC control areas. The resulting effect depends on the global state of operation of the DC grid and on the ability of the remaining converter stations to accommodate the power surplus. However, a severe case will occur regarding an operational scenario where the DC grid is highly loaded. In this case, the sudden loss of a converter results in a DC overvoltage, which needs to be mitigated in order to avoid cascading disconnection of DC grid components. Being a similar effect to the case of an AC mainland fault, similar mitigation strategies can be used in the moments subsequent to the event. However, it may require combination with the use of wind turbine pitch control systems in the case where the remaining structure of the DC grid is not able to fully accommodate the offshore wind farms' generated power.

A preliminary classification of relevant contingencies may concern the affected grid, thus distinguishing AC side contingencies (concerning offshore and onshore AC grids) from DC side contingencies. Table 4 presents a list of possible AC side contingencies (also affecting the coordination between AC grid and DC grid protection systems); the possible consequences of the contingencies and the required (preventive and/or corrective) actions to face

Table 4. Relevant AC contingencies, effects, and possible countermeasures

AC grid event	Effects on DC grid	Subsequent effects on mainland AC grid	Specific control / defence requirements
AC fault (short circuit) close to the DC grid PCC	Sudden drop of DC grid power evacuation DC grid power imbalance (excess) DC grid instability (high voltage) WF tripping / HVDC grid outage	AC grid power imbalance (deficit)	DC grid emergency controls: Fault Ride Through to AC faults (possibly including actions on offshore wind generation) Mainland AC grid: load shedding
Sudden weakening of the AC grid close to the DC grid PCC (due to loss of lines also as a result of multiple faults)	Sudden drop of power evacuation capability to the AC grid	Instability	Special schemes similar to Fault Ride Through

the problems, with particular focus on System Protection Scheme (SPS) requirements, are also reported. Table 5 is focused on DC side contingencies, stressing their effects and the possible countermeasures.

Table 5. Relevant DC contingencies, effects, and possible countermeasures

DC grid event	Effects on DC grid	Effects on mainland AC grid	Requirements
DC line (cable) fault: - pole-to-ground, - pole-to-pole (unlikely event)	DC grid power imbalance (power surplus) DC overvoltages Instability	If fault is cleared timely, DC grid remains in operation, however the AC grid is disturbed. Impact on AC also depends on the final DC grid configuration (see item below)	DC grid protection: Fast clearance DC grid emergency controls to limit power unbalance (possibly including offshore wind generation rejection)
Loss of one or more DC cables (after fault clearing)	DC power flow rearrangement Possible overloads of cables whose flow is not imposed by converters; DC grid cascading Instability, risk of DC grid complete outage	Power injection rearrangement among onshore converters, possible overloads in the AC grid, risk of cascading due to higher power injection in onshore converters In case of DC grid complete outage, large power imbalance	DC grid controls: -Corrective/preventive redispatching of power injections at GS-VSCs; -Corrective/preventive curtailment of offshore wind generation
Loss of WF-VSC	DC power flow rearrangement (overall power is reduced)	Power unbalance (deficit)	Mainland AC grid: conventional generation reserve, load shedding (in case of large power deficit)
Loss of GS-VSC	DC power flow rearrangement and possibly DC grid power imbalance (mismatch between WF generation and power delivered to onshore grid, in case the healthy onshore converters reach their maximum power in taking over the power of the lost converter) DC overvoltages Instability	Power injection rearrangement among onshore converters, possible overloads in the AC grid, risk of cascading Power unbalance (deficit)	Mainland AC grid: load shedding DC grid emergency controls to limit power unbalance (including offshore wind generation rejection)

The topic of how to make the DC grid stronger with respect to AC faults and onshore converter loss will be further dealt with in the section devoted to ancillary services. It is worth remarking here that:

- Fast and selective protection systems, aimed to clear faulted components by disconnecting them, contribute to increase stability margins;
- As a further means to allow increased exploitation of the DC grid i.e., moving the stability limits farther and farther, defence systems can be deployed, such as generation tripping (or fast reduction) schemes, energy dissipation via resistors etc. If no such systems are available, the DC grid must be operated in a less flexible, less stressed way.

5.4. Security of the AC System

The previous section remarked that in case of DC grid providing large power injections into the AC grid(s), mutual interactions must be considered in view of assessing the security of operation of the overall AC/DC grid. Indeed, disturbances on one grid may impact on the other and back on the former, and possibly lead to widespread disturbances. However, benefits for operational security may also arise from DC grids.

This section focuses on the improvements that a DC grid may bring to security of operation of AC grids. Leveraging on the flexibility of the DC grid in dispatching the offshore wind power to the onshore converters, security margins of the AC grid can be increased. In particular, the DC grid may behave as a resource to relieve AC grid overloads. This opportunity can be exploited either on a preventive stage, i.e., the action is implemented before overload occurrence (e.g., caused by a contingency), or on a corrective stage, i.e., after violation occurrence. Figure 11 reports a schematic example to explain the possible action of DC grids in relieving overloads on AC systems: in the specific operating condition, the outage of line L2 implies an overload on Line L1 (see large red arrow in Figure 11 (a)). In order to relieve the overload one can shift an amount ΔP of power from one DC injection to the other: this action brings the current on line L1 inside N-1 security limits.

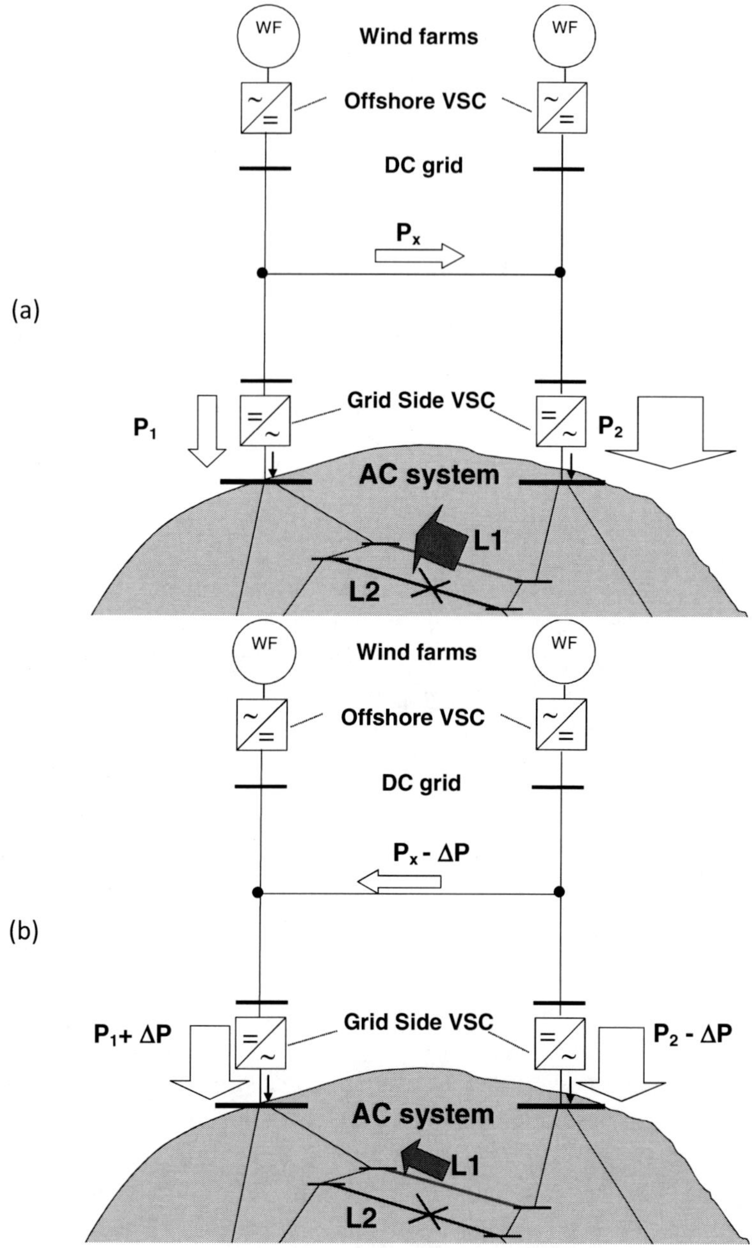

Figure 11. Power "shifting" principle deployed to relieve a branch overload in the AC system: N-1 condition (a) before redispatch, (b) after redispatch.

The DC grid could thus be regarded as a provider of ancillary services for congestion resolution, similarly to conventional generators. However, the complexity of DC grid redispatching is higher than that of individual generators. In fact, a change in one onshore injection affects the whole DC grid. In particular, if the whole offshore wind power generation has to be accommodated, any change in one onshore injection must be reallocated on the other onshore injections according to a power "shifting" principle (illustrated in Figure 11). This constraint introduces significant complexities at the technical and market levels. Of course, power injection into an onshore AC converter may be reduced by reducing the offshore wind generation, and leaving the other onshore converters unchanged. However, this solution would not typically be preferred as it would probably be more costly for the system than resorting to conventional generation redispatch.

These functions can be implemented into optimisation tools, like the ones used by TSOs for day ahead congestion management or online security control support. The added value brought by DC grids to congestion resolution is that DC injection shifting is included as a control resource in addition to conventional generation redispatch.

The objective of the envisaged control strategies in presence of DC grids is to minimize the risk of congestions/overloads, while minimizing either the redispatching costs or the redispatched power associated to conventional generation and/or power injections from the DC grid. DC power shifting is allowed, as far as security limits of the DC grid are met. The redispatching costs relevant to the different onshore DC grid converters are assumed to be defined in the Ancillary Services Market. Depending on the market design and relevant market strategies of operators, "shifting" power injections of the DC grid from one terminal to another may be less costly than redispatching conventional generators.

By applying the above control strategies, it can be shown that:

- DC grid injections can help solve congestions on AC systems;
- The higher the number of DC grid injections the higher the flexibility of DC grid in preventive actions to solve AC system congestions. This implies assuring security on AC systems at lower redispatching costs.

6. PROTECTION

A DC grid needs to be protected against short-circuits. It is very easy to detect the occurrence of a fault (voltage collapse, overcurrent). However, it is more complex to be selective and quick enough to comply with the required fault clearance time and the controllability of the converters. Both DC pole-to-pole faults and DC pole-to-ground faults should be cleared. Then, the total delay for clearing the fault (fault detection, plus selectivity algorithm time, plus breaking time of the DC Circuit Breaker - DCCB) is in the order of some milliseconds[9]. Moreover, unwanted trips of breakers should be avoided in case of a fault on another link as well as in normal operation (e.g., in case of manoeuvring actions). Finally, backup protection should handle a malfunction of a *main* protection or a breaker failure.

Only simple algorithms can fit the very short delay allowed for fault detection. Overcurrent and undervoltage criteria are robust, but not selective. This is the same for current variation and voltage variation criteria. The simplest selective algorithm is a current differential one, which is based on the comparison of currents at the ends of each device[10]. For a substation, a local differential protection can be developed easily. For a DC line, a differential protection needs fast communication (fibre optics typically) between the ends, whose disadvantage is the communication delay that can be problematic for protecting DC lines longer than 200 km (1 ms transmission delay in fibre optics to be compared with the total 2-3 ms allowed to trip the breaker). Finally, on DC links carrying series reactors, algorithms based on the comparison of voltage variation on the sides of the reactor can be imagined, regardless of the question of fitting the reactors in an evolved DC grid with breakers withstanding overcurrents.

Much R&D is ongoing on this topic [58]-[59]. In the TWENTIES project, two of these algorithms have been successfully developed, implemented and tested in a low-scale mock-up, combining real cables and converters and real-time simulation [21], [23], [26]-[30]:

---

[9] It is worth recalling that DC protection requirements are quite more challenging than AC ones in terms of both system design and equipment performances, due to the shorter time for intervention and the extreme difficulty to break DC current.

[10] A fault is detected, when the current at one terminal is different from the current at the other terminal.

- A current differential protection
- A backup protection based on current variation combined with a breaker failure detection.

6.1. Dependency on DC Circuit Breakers

In case of a point-to-point connection, there is no need for any kind of DCCB, as AC circuit breakers are used to isolate the whole DC link. Whether or not a DCCB is used leads to the same consequence for point-to-point connections, since there is no alternative path to transmit offshore power: as offshore point-to-point connections are likely to be monopolar schemes (a bipolar HVDC link would be unduly expensive considering its footprint – especially offshore, – complexity and the little benefits it may provide), a DC fault (either pole-to-pole or pole-to-ground) will result in loss of the whole point-to-point connection. In the unlikely case of a bipolar scheme with metallic return cable, the remaining transmission capacity would be respectively 50% and 0% for pole-to-ground and pole-to-pole faults.

In the situation of offshore (tree-like or meshed) DC grids, a DC short-circuit is a major contingency which would result in the loss of the whole structure in the absence of a DCCB. On the other hand, if the DC grid is duly designed, the DCCB and DC protection algorithms will be effective, i.e., the protection system can isolate the fault thus limiting its impact and stopping the rise of the fault current.One common assumption is, indeed, that the operation of a true 'DC grid' will only be possible if DC circuit breakers are developed and included in the grid.

In respect of future DC grids, if a DCCB is not used, a fault anywhere on the DC grid must be cleared by operation of circuit breakers on the AC side of all terminals. In this case, all export from the DC grid or power transfer through it will be interrupted. This will cause Offshore Wind Farms (OWFs) operating on the DC grid to shut down but will also impact on the AC power systems receiving power from the DC grid. However, these AC systems are already operated with the expectation that some level of 'loss of infeed' will be experienced from time to time and some amount of primary reserve will be carried to ensure that the AC system frequency does not go outside certain limits as a consequence of a loss of infeed event. Thus, the total power exported by a DC grid to any AC system synchronous area via one or more terminals should be restricted to the level of the primary reserve of that area. The level of reserve should be determined by a cost-benefit analysis comparing the cost of reducing the level of loss of infeed risk with that of carrying additional reserve.

Reverse blocking converters are a feasible extension to existing multi-level converters. This could allow rapid interruption of the current from the AC side and clearing of the DC fault under zero voltage/current. Interruption of the network could be of short duration and OWFs may be able to ride through. Reverse blocking converters can introduce additional losses over existing multi-level converters.

If the benefits of a meshed DC grid (as distinct from one that is not meshed) are to be realised in terms of continuity of supply following faults in certain locations, DCCBs and sufficiently accurate fault identification and location are essential. However, it is also the case that the performance required of a DCCB depends on where a fault occurs, on the design both of the grid and the converters at each terminal, and on operating conditions on both onshore and offshore side.

DCCB technology is still at its early stages. Prototype equipment has been reported by two major manufacturers [30],[60]-[61]. A feasibility test was performed within the TWENTIES project, consisting in the interruption of a sinusoidal current well before its natural current zero under a voltage of few tens of kV (medium voltage range). The rate of rise of current was 3.2A/μs. Similar interruption tests were also performed at voltages exceeding 100 kV in the lower part of the voltage range used for long distance power transmission. The demonstrator is designed for operation at a rated voltage of 120 kV, but during interruption the voltage rises transiently to approximately 180 kV. Details can be found in [29]-[30].

6.2. Limits to Wind Farm Capacity on Single, Contiguous DC Grids

The phrase 'single, contiguous DC grid' is used to mean a DC power island that is electrically continuous before the occurrence of any fault and the deployment of any consequential action. (The different DC electrical islands might be interconnected but only via an AC system through at least two converters.)).

If it is assumed that DCCBs will not be used in the point-to-point connection case, the maximum power that could be generated on a single electrical island will be restricted by the ratings of individual cables and converters and by the maximum loss of infeed that could be survived by the

AC system to which it is connected. The maximum loss of infeed equals the primary reserve carried on the AC system. The precise level depends on a cost-benefit analysis between the cost of additional primary reserve and the benefit of allowing a higher loss of infeed.

The above considerations set the limit on the maximum power that could be generated on a single, contiguous DC grid. However, the installed OWF capacity could be greater than that level, accounting for expected unavailability of individual turbines, wake effects or the likely diversity of output (in case of quite well spatially separated OWFs connected within a single cluster).

An example of a DC grid that does not have DCCBs is shown in Figure 12. Each converter is assumed to have a capacity of 1 GW. In the shown operating condition, 3 GW of power is being generated offshore; 2 GW of that power is being directed to AC synchronous area 1 and 1 GW to AC synchronous area 2.

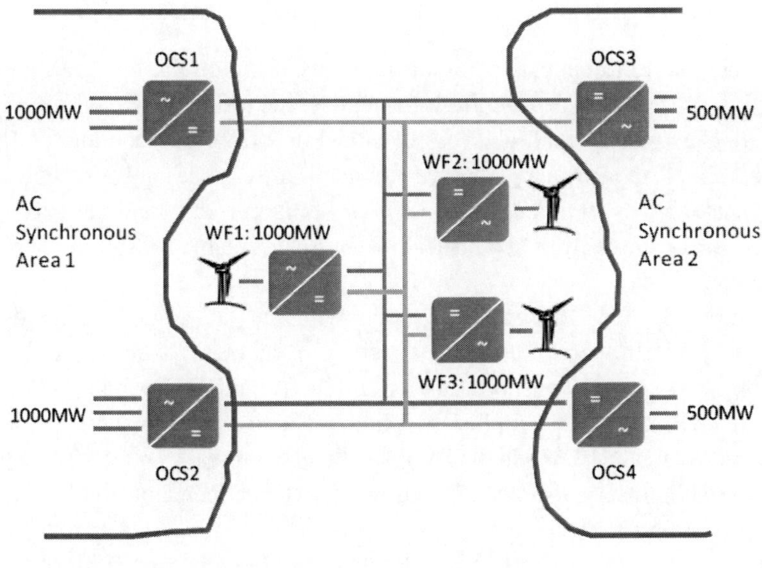

Figure 12. Example DC grid and dispatch of power totaling 2GW to area 1 and 1GW to area 2.

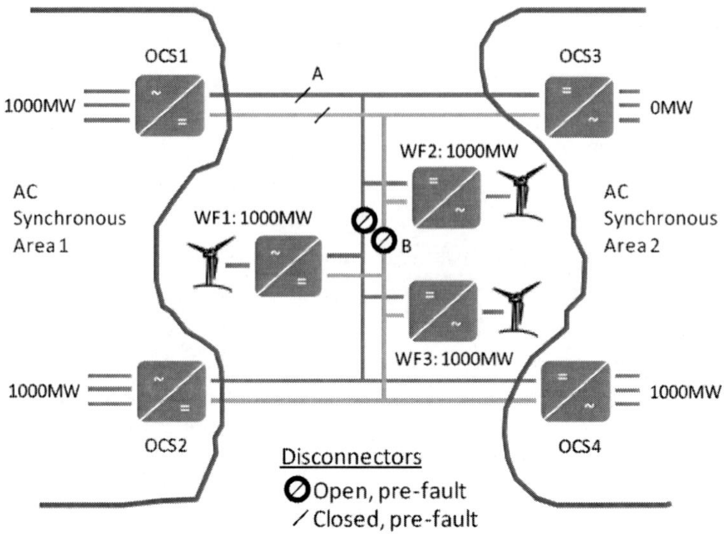

Figure 13. Example DC grid with a pre-fault split introduced at location B.

If there were to be a short-circuit fault anywhere on the DC grid, without DC breakers, the passage of current from all of the terminals would need to be blocked. This would result in a loss of infeed of 2 GW to area 1 and of 1 GW to area 2. If the primary reserve in each area were only sufficient to compensate a loss of infeed of 1.32 GW (as is currently the case in Great Britain, for example), this would lead to an unacceptable deviation of system frequency in area 1.

Figure 13 shows the same DC grid with the DC grid configured in a different way. In particular, disconnectors have been opened pre-fault at location B to separate onshore converters OCS1 and OSC3 plus WF2 from converters OCS2 and OCS4 plus WF1 and WF3. In order to generate 1 GW at WF2, and to send 2 GW and 1 GW respectively to area 1 and 2, the dispatch of power should be as shown. In this way, two independent contiguous DC grids have been formed.

If one considers a short circuit fault on the branch connecting OCS1 to location A, only currents through OCS1 and OCS3 need to be blocked onshore and surplus power from WF2 dissipated. The power flow through OCS2 and OCS4 can continue. This results in a loss of infeed to area 1 of only 1GW (within a Great Britain limit of 1.32GW). Provided the location of the fault can be correctly identified, disconnectors can be opened at location A as in Figure 14, and, assuming that the loss of infeed limit in area 2 is at least 2 GW, the disconnectors at location B can be closed to provide flexibility in the export of power from WF2. OCS3 can be returned to service and WF2 restarted. (In this case, the total power being generated offshore equals the total remaining onshore converter capacity, thus the disconnectors at location B could have been left open).

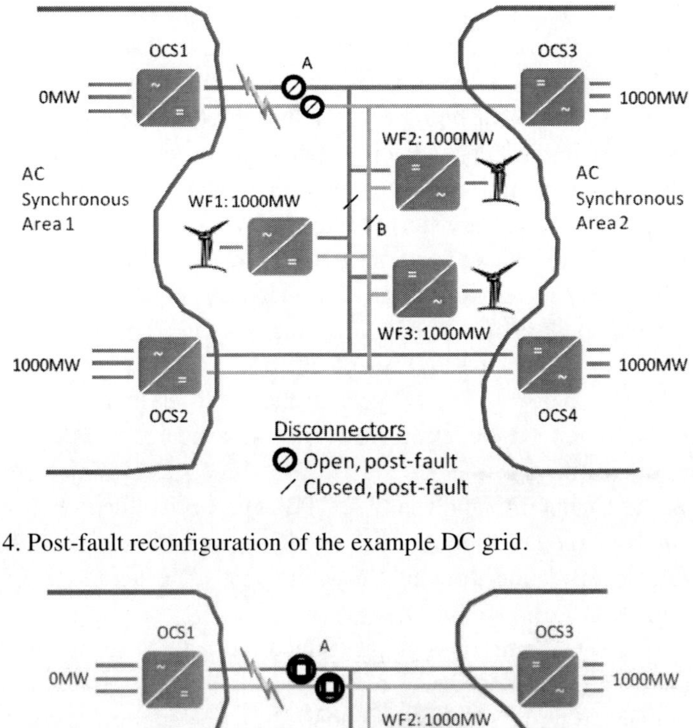

Figure 14. Post-fault reconfiguration of the example DC grid.

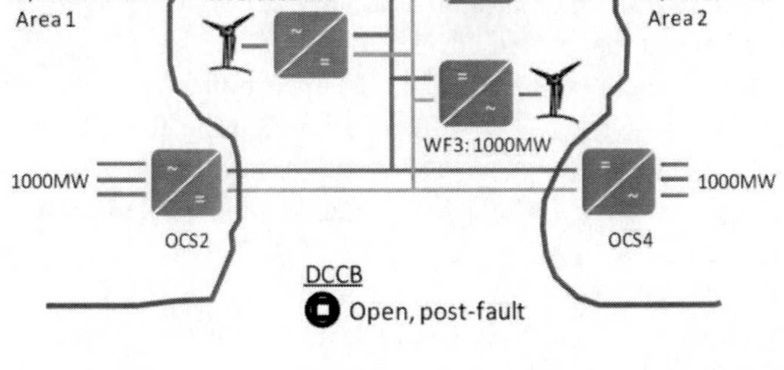

Figure 15. Example DC grid with DCCBs at location A that succeed in clearing a fault between OSC1 and location A.

Figure 15 shows the same system with DCCBs installed. In this case, provided DC protection correctly identifies the location of the fault and the DCCBs operate successfully, it would not have been necessary to partition the DC grid pre-fault. Only OCS1 would be disconnected and the power from WF2 could flow continuously, without interruption. However, the value of this continuous operation (equating to the cost of the interruption in the case without DCCBs) should be compared with the cost of the DCCBs.

The main benefit of a meshed DC grid is the availability of a parallel path in the event of a fault on one branch. This would allow to continuously transfer at least some power but only if faults can be located and isolated on the DC side. This, in turn, depends on installation and successful operation of DCCBs. The limit to operation is then determined by the power flow condition (how much power is being transferred into each AC system to which the DC grid is connected), the location of the DCCBs and the impact of any fault outage. This impact in terms of loss of power transfer into any AC system must be less than the loss of infeed limit for that system.

For a DC grid, again the loss of infeed limit relative to primary reserve applies. However, here, if the DC grid connects different AC synchronous areas, the loss of infeed limit for each injection should be referred to the primary reserve of the specific AC area to which the specific injection refers. In addition, the availability and use of DCCBs makes a difference. If they are not used, faults on the DC grid must be cleared by operation of circuit breakers on the AC side of all the terminals meaning that all generation and power transfer from the DC grid are lost.

This clearly sets the limit on the net infeed into any AC synchronous area to which the DC grid is connected.

Notice that the term ‘single contiguous DC grid’ applies to the pre-fault operational state of the HVDC facilities and the OWFs to which they are connected. Aside from when a high continuity of service is required, there is no strict requirement that all the DC network branches must be interconnected pre-fault. That is, there could be normally open points at particular hubs or substations. This would have the effect of configuring the offshore network facilities into multiple contiguous DC grids. A fault on any of them would cause only that particular DC electrical island to be isolated by operation of AC side breakers.

A particular loss of infeed limit could thus be readily managed by configuring the DC grid into a number of sections that are only interconnected, pre-fault, via the AC side such that no section exports more power into any AC system location than the loss of infeed limit for that AC system. Should a fault occur and any cable, converter or bus section be lost from service, the total power transfer capability is likely to be reduced but, depending on the design of the DC grid, the operational state of the network facilities can be reconfigured and most, if not all, of the power transfer resumed after only a relatively short interruption.

6.3. Impact of DC Grid Design on DC Breaker Stresses

In the absence of a DC breaker, the major consequence of a DC short-circuit is the dramatic rise of fault current through the DC grid in a very short period of time after the fault. This results from various phenomena, namely the discharge of the DC cables (hence the influence of the cable length, the DC grid meshing and the value of DC inductances have to be taken into due account), and the current feeding the fault from onshore and offshore sources; the latter, in particular, depends on the converter topology (with or without current blocking capability), considering various smoothing reactor values, and on the strength of the various AC networks (which include the offshore wind farms). All those parameters affect the severity of a DC fault current:

- Large networks will increase the energy stored in cable capacitance which will discharge into the DC fault.
- The presence of multiple converter stations will provide increased AC infeed which may be concentrated in localised sections of the network.
- Highly meshed networks will lower DC network impedance leading to faster DC voltage collapse and larger line discharge current.

The location of DCCBs will also affect the stress they may be submitted to: a DCCB sited close to a converter station will be subject to fault currents similar to those of the converter. These currents are predominantly defined by the design of the converter – typically the internal arm inductance and the inductance of the coupling transformer. The DC grid design therefore has limited impact on the stress seen by the converters and DCCBs connected at their ends.

On the contrary, DCCBs located at internal nodes will experience the cable capacitance discharge current and the AC contribution to DC fault current from multiple converters. Therefore, the design of the grid will determine the energy that must be absorbed by the DCCB when interrupting the DC fault current.

Considering different sets of the previously described parameters, significant DC fault categories have been identified, as recalled in Table 6.

The requirements of a DCCB lie in categories I and II, with proper assumptions on the performance of a DC fault detection algorithm and considering cable length up to 200 km.

Table 6. DC fault categories for DCCB performance [27]

DC fault category	Fault current (amplitude and peak time after fault inception)
Category I	5 kA after 15 ms
Category II	45 kA after 10 ms up to 50 kA after 10 ms
Category III	80 kA after 4 ms, 55 kA after 2.5 ms up to 65 kA after 2 ms

6.4. AC Protection Issues

Onshore converters in a DC grid, being of the VSC type, can control AC voltage independently of the active power. Protection miscoordination of existing AC protections in the presence of VSC converters may become an issue [63].

In fact, when an AC transmission line close to a PCC undergoes a Short Circuit (SC) fault, the PCC voltage is reduced. This in turn affects voltages and currents of the AC transmission grid close to the PCC. Since the capacity of VSC stations is generally large for the offshore power transmission, their impact on the line voltages and currents will be significant. In particular, a significant impact on the performance of existing protection schemes on the AC grid, such as distance protections, is expected. The basic principle of distance relaying is to measure the apparent impedance using the voltage and current viewed by a relay, which approximately determines the distance between the relay location and the fault point during a SC fault. The apparent impedance is compared with pre-set relay operation characteristics to decide whether the fault is within the protected "zone". Appropriate time delays are used to allow primary and back-up functions among the distance relays. The Zone 2 relay is a backup protection function that is usually located on the remote end of a line adjacent to the protected line.

Figure 16 shows the PCC (named PCC2) of one onshore converter (GS-VSC2) and two AC lines connected to the PCC. When line j-k experiences a three-phase SC fault F, it is viewed by relay B as a fault in Zone 1. Assuming the fault impedance is zero, the impedance viewed by relay B is the line impedance from the fault point to the relay location. In case of a failure of Zone 1 relay, relay B should operate: if it does not, the isolation of fault F will rely on its backup protection i.e., relay A that is located on the adjacent line i-j.

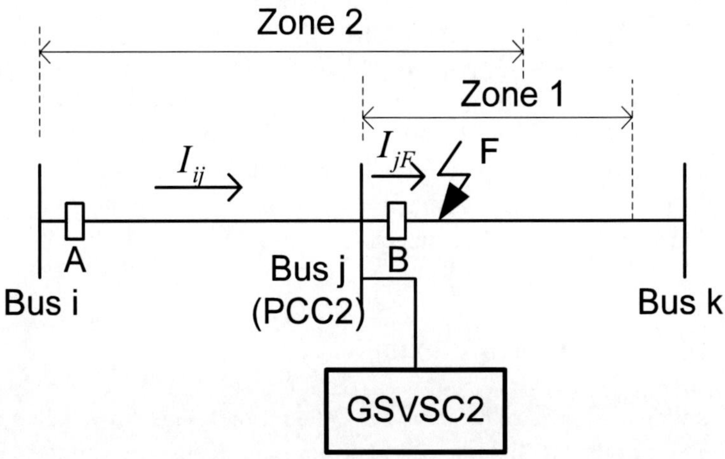

Figure 16. Portion of an AC grid with a converter and two relays.

Compared with the excitation systems of synchronous generators, the converter reactive power regulation demonstrates a much faster dynamic response. During fault F, GS-VSC2 is triggered to provide reactive power support to boost the reduced PCC2 voltage. The resulting PCC2 voltage increase leads to a reduced current on line i-j. As a result of the impact of VSC fast control (in the order of milliseconds) on line voltages and currents, the fault distance of the SC fault can be overestimated by its backup relay located on the adjacent line. It is possible for a Zone 2 fault to be viewed as a Zone 3 event, resulting in mis-coordination between protective relays. If the delayed fault clearance exceeds the critical clearing time (CCT), the system can suffer instability and potentially trigger a sequence of cascading events.

Adaptive distance relays have been introduced to overcome this problem, in order to enhance the overall reliability of the protection scheme.

7. ANCILLARY SERVICES

There are many important ancillary services traditionally provided by conventional thermal or hydro-based generation units such as voltage and frequency control[11]. Additionally, conventional generation units intrinsically provide inertia to the system, which is a fundamental characteristic in order to assure its stability. The large-scale integration of wind power naturally displaces at least some conventional generation units, thus affecting ancillary services provision and global system security [64].

A fully operational HVDC grid will play a key role in order to create larger interconnections among AC systems and to integrate offshore wind farms. The importance of such infrastructure requires its active contribution for ancillary services, namely: (1) fault ride-through (FRT) capability in case of mainland AC faults, (2) primary reserve and inertia emulation and (3) contribution to damping of electromechanical oscillations.

7.1. Fault Ride-Through Capability

The anticipated level of offshore WF integration brings new challenges for safe operation of the AC mainland grid. Taking into consideration the security of operation of onshore AC grids and the amount of offshore wind power the DC grid injects into the onshore AC systems, it is anticipated that AC-side fault ride-through (FRT) functionalities will be also required for DC grids. Therefore, the identification and implementation of control strategies to provide AC-side FRT capability in DC systems is a key issue.The Fault Ride-Through (FRT) requirement can be generally defined as the capability that a given generation unit has to remain connected to the grid in case of a voltage sag which is usually related with faults occurring in the grid. This issue started to be a matter of concern following the large scale installation of small-size distributed generation units, such as wind generators. Since different wind power generation technologies were not able to remain connected in case of voltage sags, large scale integration of such units might compromise system security. In fact, large scale penetration of generation units without the

[11] Ancillary services may be classified in different ways. According to [62], ancillary services can be grouped into the following: 1. frequency control, 2. voltage control, 3. spinning reserve, 4. standing reserve, 5. black start capability, 6. remote automatic generation control, 7. grid loss compensation, 8. emergency control action.

capability of remaining connected during low voltage periods leads to significant amounts of generation tripping in case of severe network faults. Such events compromise system security at the generation/transmission level, precluding the desirable increase of non-conventional generation units such as wind power.

When applied to the HVDC grid with offshore wind farms, the FRT capability can be seen as the ability of DC converter associated with the affected AC area to remain connected to the grid during low voltage periods. The FRT provision can be attained by VSC that are used to establish DC grids. However, proper control strategies must be defined to guarantee that a converter's maximum current is not exceeded during the fault occurrence [65]-[69].

A fault in the onshore AC system, considering the current limits of VSC stations, will cause a significant drop in the power delivered by the converter. Therefore, the power not being delivered to the onshore AC grid will cause significant over voltages in the DC grid. Indeed, under these circumstances, the resulting power imbalance charges the DC circuit capacitors, resulting in a fast increasing DC voltage which may destroy the HVDC equipment if no countermeasures are adopted. Therefore, the maximum time to reduce power in order to prevent over voltages in the DC capacitor banks plays a key role regarding the control strategies to be developed in which the DC voltage control should be as fast and tight as possible. Hence, advanced control functionalities are proposed in the control systems of the local VSC-HVDC stations aiming at an effective active power accommodation that can be achieved following two main possible approaches or a combination of them:

- The first approach consists in installing a DC chopper resistor on the DC side of each onshore VSC-HVDC.
- The second approach consists in endowing onshore and offshore VSC with coordinated cascaded local control rules allowing mimicking AC onshore voltage profile reduction into the AC offshore voltage profile. This strategy rationale allows offshore wind generators to locally accommodate/dissipate active power not being delivered to onshore grids. Additionally, this strategy also allows off-shore wind farms to ride though faults occurring in their own off-shore AC grid. In particular, this approach can be adopted to allow the DC grid go through loss of onshore grid converters, as the involved phenomena (lack to evacuate power to the SC grid) are very similar.

It is worth pointing out that these approaches do not rely on communication systems, whose requirements (speed, reliability) would otherwise be very stringent. In fact, the fault ride-through capability can be attained by different control philosophies depending on the Wind Energy Converter (WEC) technology. Nevertheless, the methodologies share a common aspect which is related with the need of accommodating the power which the generator cannot inject into the grid due to the voltage dip in order to guarantee the generator does not lose stability.

As an illustrative example, Figure 17 shows the time to reach a DC-link overvoltage equal to 120% of the nominal voltage, as a function of the power not being delivered to the AC grid (active power drop) in case of a 400 MW VSC-HVDC converter station with 75 μF DC capacitor. Therefore, Figure 17 allows to assess the maximum time to perform the active power accommodation/dissipation in order to avoid the DC voltage to overpass the maximum admissible value, which can be as low as 10 ms in case of severe faults leading to almost zero active power injection into the AC grid.

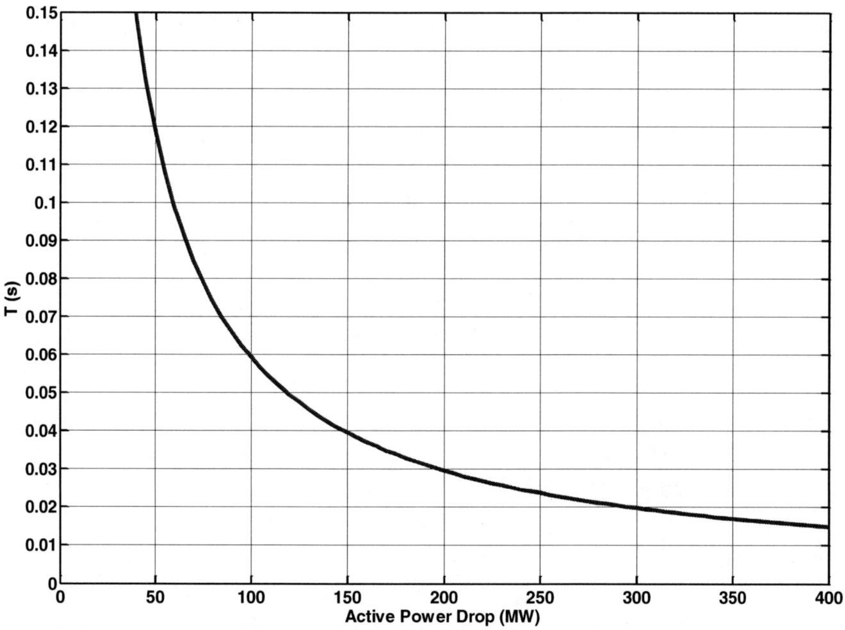

Figure 17. Time to reach maximum DC voltage as a function of the non-delivered active power.

7.2. Frequency Regulation

Large-scale integration of wind energy leads to a displacement of conventional generation units that negatively impacts the behaviour of grid frequency (increases the rate of change of frequency and the absolute frequency deviation) in the aftermath of disturbances, thus affecting the load-generation balance. To mitigate such bottlenecks the possibility of using wind generators to provide primary frequency support and inertia emulation is being requested in some grid codes. The predicted massive integration of offshore WF contributes to increasing frequency stability related problems in AC systems. Despite being desirable to make offshore WF connected through HVDC grids to contribute towards frequency regulation in AC systems, the DC link connections fully decouple AC areas and offshore stations. Both cost and reliability issues will preclude the development of high speed communication and control centres that should be able to process and communicate real-time information, regarding the core application for frequency control purposes. Additionally, the event of a delay or communication failure can jeopardize the operation of a DC network with primary frequency control capability under the solicitation of a disturbed AC system.

In order to overcome the aforementioned difficulties, an innovative approach consists in the identification and development of local controllers to be installed at HVDC-VSC which will autonomously allow the provision of frequency control services. The proposed control strategies make use of a cascading reproduction of AC grid frequency deviations into DC grid voltage variations. Subsequently, DC grid voltage variations are used by the offshore HVDC-VSC for controlling the offshore WF AC grid frequency. Thus, offshore WF AC grid frequency variations will be the driving signal for frequency regulation loops to be adopted at the wind generator level.

Further details are now provided about the abovementioned cascaded control strategies.

In particular, the control characteristics of the onshore converters (see Figure 18) refer to two operation modes:

- Normal Operation Mode: the load and generation in the onshore AC grid to which the converter station is connected are balanced, thus resulting in very small frequency deviations. In this situation the converter is controlled by the previously presented k_{PV} (DC voltage /

Active power) droop. Thus the active power flow will depend on the existing DC voltage profile.

- Disturbed Operation Mode: the AC grid frequency to which the VSC-HVDC converter station is connected drops below / rises above a certain margin due to any load/generation imbalance. In these situations, primary frequency regulation services are requested to be autonomously deployed in order to help stabilizing mainland AC grid frequency.

Regarding the mode of deploying the offshore WF reserves, it firstly requires an additional control mechanism that translates onshore AC grid frequency grid variations to DC grid voltage profile variations. In this case, the onshore VSC-HVDC converter station measures the terminal frequency and whenever it drops below / rises above the referred margin, the actual operation point defined by the active power/DC voltage droop control mechanism enters in standby and a new droop control mechanism relating onshore AC grid frequency and DC grid voltage enters in operation, according to the following equation:

$$V_{DC} = V_{DC}^{0} - k_{fv} \times \Delta f$$

where V_{DC} is the reference value for the DC voltage at the onshore VSC-HVDC, V_{DC}^{0} is the pre-disturbance DC voltage resulting from the operation of the converter in the normal mode governed by the k_{PV} droop, k_{fv} is the frequency/DC voltage droop and Δf is the onshore AC grid frequency deviation.

Regarding system operation in the moments subsequent to a frequency deviation in the mainland AC grid, it is necessary to define the mainland AC grid frequency dead-band (as depicted in Figure 18) that requires the deployment of reserves from off-shore wind farms. In fact, off-shore wind farms are not suitable to be continuously asked for adjustments regarding a response to frequency stabilization issues due to the stress induced over the turbine mechanical components. Therefore, this issue justifies the need of defining a frequency dead-band where off-shore wind farms are not requested to participate in the frequency regulation process. As an illustrative example, and regarding the European Network of Transmission System Operators for

Electricity (ENTSO-E) operation rules, it will be possible to suggest that offshore wind farms will be able to provide primary frequency control for frequency deviations greater than ±20 mHz (conversely, conventional power plants continuously provide primary frequency regulation services). However, a precise definition of such dead-band greatly depends on the characteristics of the frequency regulation capabilities available in AC mainland systems.

In the event of a frequency deviation in a mainland AC grid larger than the pre-defined dead-band, onshore VSC-HVDC should switch from the Normal Operation Mode to the Disturbed Operation Mode. For onshore AC grid frequency deviations smaller than the assumed dead-band, the converter should operate inside the dashed zone (see Figure 18), over the k_{PV} droop line. The solid grey zone shown in Figure 18 defines a sliding window corresponding to the onshore converters operational zone when the k_{PV} droop mechanism enters in standby and the k_{fv} droop mechanism is initiated. This moving window is continuously adapted to the converter DC voltage operation point in order to avoid sudden voltage variations when changing between the previously referred operational modes. For frequency deviations greater than $\pm$ 20 mHz (illustrative value) the k_{fv} droop will be responsible for defining the DC grid voltage operational point at the DC side of the onshore VSC-HVDC. When recovering from the onshore AC grid disturbance, the converter will smoothly commutate from the disturbed operational mode to the normal operational mode. Thus, the k_{PV} droop will assume the converter control.

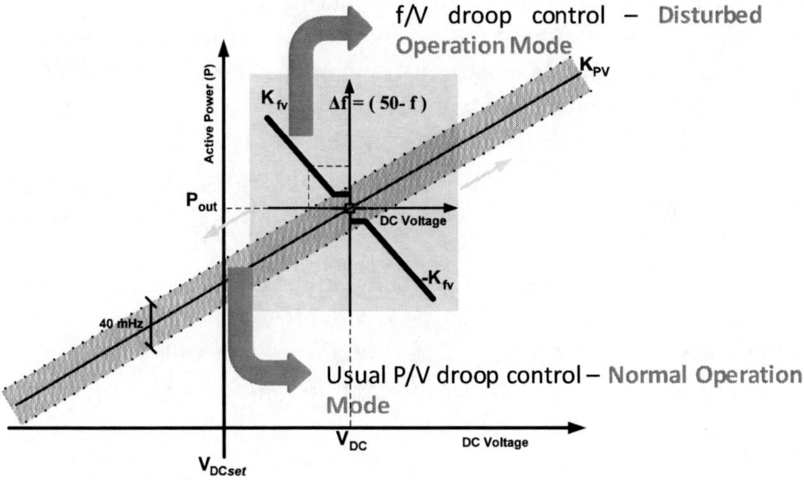

Figure 18. Droop control mechanisms for VSC-HVDC onshore converter stations.

7.3. Oscillation Damping

The displacement of conventional power plants to accommodate the surplus of wind generation can lead to loss of supplementary damping usually provided by synchronous generators, if they are equipped with power system stabilisers (PSS). Within these circumstances, critical damping levels of inter-area modes of oscillation will occur. Therefore, performing small signal stability analysis of interconnected DC/AC systems has become an important task. Moreover, although VSC-HVDC power stations do not participate in the electromechanical modes of oscillation, they can provide additional damping by means of PSS based controllers installed in the active power control loop that are able to modulate active power flows though the DC grid in order to mitigate the effect of oscillatory modes.

One frequent cause of instability is the lack of damping of the so called electromechanical modes of oscillation, which are related with low frequency (0.1-2Hz) power oscillations that occur among the rotors of synchronous machines, essentially due to some controller settings and due to large power flows through weak transmission lines. It has been demonstrated that wind generation has some impacts on the damping levels of a power system, and that the nature of these impacts depends on factors like wind generation conversion technology or network configuration. Depending on the operating conditions, the integration of wind power may then lead to an improvement or reduction of these damping levels. A situation of reduced damping may occur when the connection of wind power in one area of the grid, replacing conventional production located in another area, increases the power flows through weak interconnection lines. This may take place in the future, for example, in the north of Europe with the construction of large offshore wind farms. The existence of a large surplus of generation in some areas may in these cases lead to increasing exchanges of power between weakly interconnected grids, giving rise to insufficiently damped oscillations between control areas.

Power system stabilizers (PSS) installed in the excitation of synchronous generators (SG) is still one of the most cost effective solutions capable of providing supplementary damping. However, in a system with a large share of wind power production these devices may no longer be capable of providing the necessary additional damping.

The DC/AC electronic converters of off-shore wind farms may be used to introduce damping in a power system since, as synchronous generators, these devices can act as power amplifiers of the auxiliary controllers responsible for adding damping to the electromechanical modes of oscillation. Nevertheless that is not straightforward because these electronic devices are not directly involved in the electromechanical oscillations.

A new Power System Stabilizer (PSS) installed in the DC/AC converter will only be effective to modify a mode of oscillation if that mode is controllable through the injection of power in the bus to which the injection itself refers and, at the same time, the mode is observable in the input signal of the stabilizer. The effectiveness of such a PSS, and consequently the contribution of the converter where it will be installed, to damp the modes of oscillation depends on an appropriate choice of the input control signals, architecture of the stabilizer and its domain of actuation regarding the control loops of the DC/AC converter.

It is also well known that the contribution of a PSS to the damping of the electromechanical modes depends significantly of the operating conditions. A solution that provides satisfactory damping for some situations may not lead to adequate results for other scenarios. For this reason it is desirable to find, for each PSS, a set of parameter values that are capable of producing the desired damping level for all scenarios or, if this is not possible, the closest level achievable for all the operating scenarios of interest. A solution with such characteristics is called a robust solution. In the case of PSS to be installed in DC/AC converters, robust solutions with the characteristics described before will also be required. Since these devices add damping in an indirect way, obtaining a robust solution may be quite difficult to achieve.

7.4. Contribution to System Restoration

The high level penetration of offshore wind power can dramatically increase loading of the transmission grid. The reduced system stability margin increases the vulnerability of a power system during severe contingencies. This may result in widespread power outages in integrated AC/DC systems following a major disturbance. System restoration after wide-area power outages can be difficult and time-consuming. Generally, the power plant selected as a black start unit is equipped with small diesel generators that provide its black start power support. For generators with steam turbines, the required black start power support can be up to 10% of the capacity of generators for boiler feed-water pumps, boiler forced-draft combustion air blowers, and fuel preparation. It is costly to provide such a large standby capacity at a power station, especially for the oversized diesel generators. With the innovation of wind technologies, the variable-speed wind turbine technology enables the regulation of power factor by absorbing or producing reactive power.

Compared with the traditional restoration equipment, wind energy can be a valuable asset to significantly lower investment and maintenance cost. By developing suitable controls and procedures, wind farms may serve as black-start units for AC system restoration through HVDC links. Flexible HVDC control can effectively alleviate transient and steady-state overvoltages and provide frequency control before the synchronization of generators. This will help to reduce the restoration time and smooth the restoration process.

Conclusion

This section reviewed some technical issues and opportunities related to the realisation of offshore DC grids aimed at offshore wind power transfer and interconnection. Large economic effort is involved in offshore projects, moreover technical barriers still exist which prevent from the implementation of meshed DC grid. However, technology is progressing and this section provided several insights on key design options, such as grid topology and converter configuration, and possible ways to operate the DC grid in a fully integrated way with respect to AC grids. A step-by-step approach has been envisaged towards DC grids, starting from a small backbone-shaped one that can be readily constructed and extended with currently available technologies, without specific equipment or systems like DCCBs or master DC grid controllers. An intermediate stage with simple meshed networks could then be achieved, for which specific equipment or systems such as a master DC grid controller would be required in addition to controls for the backbone structures. A final stage with complex meshed DC grids would require to completely solve issues regarding, for example, protection systems, which are currently being addressed by R&D departments of manufacturers and by research institutions.

References

[1] European Wind Energy Association EWEA. *Wind energy scenarios for 2020*; 2014. www.ewea.org.

[2] European Wind Energy Association EWEA. *Pure power: Wind Energy targets for 2020 and 2030*; 2011. www.ewea.org.

[3] European Commission. *A Roadmap for moving to a competitive low carbon economy in 2050*; COM (2011) 112. 2011. http://eur-lex.europa.eu/.

[4] e-Highway2050 EU FP7 Project. www.e-highway2050.eu/e-highway2050/.

[5] European Commission. *Offshore Electricity Grid Infrastructure in Europe*; 2011. www.offshoregrid.eu/.

[6] *European wind integration study (EWIS): Final Report;* 2010. http://www.wind-integration.eu/.

[7] ENTSO-E. *Offshore Grid Development in the North Seas: ENTSO-E views*; 2011. http://www.entsoe.eu.

[8] ENTSO-E. *Ten-Year Network Development Plan (TYNDP)*; 2014. http://www.entsoe.eu.

[9] ENTSO-E. *Scenario outlook and adequacy forecast (SO&AF) 2014-2030*. 2014. http://www.entsoe.eu

[10] Kriegers Flak Offshore Wind Farm. 2014. http://www.energinet.dk/EN/ANLAEG-OG-PROJEKTER/Anlaegsprojekter-el/Kriegers-Flak-havmoellepark/Sider/default.aspx (accessed March 18, 2015).

[11] COBRAcable Project. 2015. http://www.tennet.eu/nl/grid-projects/international-projects/cobracable.html (accessed March 18, 2015).

[12] Moray Firth Project. 2014. http://morayoffshorerenewables.com/.

[13] North Seas Countries' Offshore Grid Initiative. *North Seas Countries' Offshore Grid Initiative: Initial Finding*; 2012. http://www.benelux.int/nl/kernthemas/energie/nscogi-2012-report/ (accessed March 18, 2015).

[14] Eirgrid. *Offshore Grid Study: Analysis of the Appropriate Architecture of an Irish Offshore Network*; 2011. http://www.eirgrid.com/media/EirGrid%20Offshore%20Grid%20Study.pdf (accessed March 18, 2015).

[15] THINK. *Offshore Grids: Towards a least Regret EU Policy. Topic 5, Final Report*; 2012. http://www.eui.eu/Projects/THINK/Documents/Thinktopic/THINKTopic5.pdf (accessed March 18, 2015).

[16] Cole, S.; Martinot, P.; Rapoport, S.; Papaefthymiou, G.; Gori, V. *Study of the benefits of a meshed offshore grid in Northern seas region.* Final Report for the European Commission; 2014. https://ec.europa.eu/energy/ (accessed March 18, 2015).

[17] ENTSO-E. *Offshore Transmission Technology*; 2012. http://www.entsoe.eu.

[18] European Union's Third Energy Package. http://ec.europa.eu/energy/en/topics/markets-and-consumers/market-legislation (accessed March 18, 2015).

[19] Directive 2009/72/EC of the European Parliament and of the Council of 13 July 2009 concerning common rules for the internal market in electricity and repealing Directive 2003/54/EC. 2009. http://eur-lex.europa.eu/homepage.html.

[20] TWENTIES EU FP7 Project. http://www.twenties-project.eu/ (accessed March 18, 2015).

[21] Despouys, O.; et al. *TWENTIES: Conclusions of a major R&D Demonstration Project on Offshore DC Grids*, CIGRE Session, Paris, 2014; Paper B4-107.

[22] Ciapessoni, E.; et al. *Impact of multi-terminal HVDC grids on AC system stability and operation*, CIGRE Session, Paris, 2014; Paper C4-106.

[23] Liu, C.-C.; et al. *Preliminary Analysis of HVDC Networks for Off-Shore Wind Farms and Their Coordinated Protection.* TWENTIES Project Deliverable D5.1; 2011. http://www.twenties-project.eu.

[24] Ciapessoni, E.; et al. *Principles of an optimal operation of the AC/DC interconnected power systems: Specification of the requirements for the HVDC grid (DCG) injections to the AC grid.* TWENTIES Project Deliverable D5.2b; 2011. http://www.twenties-project.eu.

[25] Denis, A. M.; et al. *Advanced results of control and protection of DC networks: behaviour, optimisation and grid extension.* TWENTIES Project Deliverable D5.3b; 2013. http://www.twenties-project.eu.

[26] Denis, A. M.; et al. *DC grids: motivation, feasibility and outstanding issues.* TWENTIES Project Deliverable D5.4; 2013. http://www.twenties-project.eu.

[27] Nguefeu, S.; Rault, P.; Grieshaber, W.; Hassan, F. *Requirement Specifications.* TWENTIES Project Deliverable D11.1; 2012. http://www.twenties-project.eu.

[28] Curis, J.-B.; et al. *Description of a Protection Plan for DC Networks: Preliminary Results towards the real-time experimentation of a small-scale model of a DC network and the 120 kV DC breaker prototype tests.* TWENTIES Project Deliverable D11.2; 2012. http://www.twenties-project.eu

[29] Curis, J.-B.; et al. *Testing results from DC network mock-up and DC breaker prototype.* TWENTIES Project Deliverable D11.3; 2013. http://www.twenties-project.eu.

[30] Grieshaber, W.; Penache, D.-L. *Test results of DC Breaker Demonstrator.* TWENTIES Project Addendum to Deliverable D11.3; 2014. http://www.twenties-project.eu.

[31] Cole, S.; Belmans, R. Transmission of bulk power. *IEEE IND. ELECTRON. MAG.* **2009**, *3*, 19-24.

[32] Flourentzou, N.; Agelidis, V.G.; Demetriades, G.D. VSC-Based HVDC Power Transmission Systems: An Overview. *IEEE Trans Power Electron.* **2009**, *24*, 592-602.

[33] Fleeman, J. A.; Gutman, R.; Heyeck, M.; Bahrman, M.; Normark, B. EHV AC and HVDC transmission working together to integrate renewable power. In *Integration of Wide-Scale Renewable Resources Into the Power Delivery System* CIGRE/IEEE PES Joint Symposium, Calgary, Canada, July 29-31, 2009.

[34] Franck, C. M. HVDC Circuit Breakers: A Review Identifying Future Research Needs. *IEEE T POWER DELIVER.* **2011**, *26*, 998-1007.

[35] Weixing, L.; Boon-Teck, O. DC overvoltage control during loss of converter in multiterminal voltage-source converter-based HVDC (M-VSC-HVDC). *IEEE T POWER DELIVER.* **2003**, *18*, 915-920.

[36] Gole, A. M.; Meisingset, M. Capacitor commutated converters for long-cable HVDC transmission. *POWER ENG J.* **2002**, ***16***, 129-134.

[37] Lidong, Z.; Harnefors, L.; Nee, H.-P. Interconnection of Two Very Weak AC Systems by VSC-HVDC Links Using Power-Synchronization Control. *IEEE T POWER SYST.* **2011**, *26*, 344-355.

[38] Jowder, F. A. R.; Boon Teck, O. VSC-HVDC station with SSSC characteristics. *IEEE T POWER ELECTR.* **2004**, *19*, 1053-1059.

[39] Du, C.; Agneholm, E.; Olsson, G. Use of VSC-HVDC for Industrial Systems Having Onsite Generation With Frequency Control. *IEEE T POWER DELIVER.* **2008**, *23*, 2233-2240.

[40] Hui, D.; Yi Zhang; Gole, A.M.; Woodford, D.A.; Min Xiao Han; Xiang Ning Xiao. Analysis of Coupling Effects on Overhead VSC-HVDC Transmission Lines From AC Lines With Shared Right of Way. *IEEE T POWER DELIVER.* **2010**, *25*, 2976-2986.

[41] de la Villa Jaen, A.; Acha, E.; Exposito, A.G. Voltage Source Converter Modeling for Power System State Estimation: STATCOM and VSC-HVDC. *IEEE T POWER SYST.* **2008**, *23*, 1552-1559.

[42] Weixing, L.; Boon-Teck, O. Premium quality power park based on multi-terminal HVDC. *IEEE T POWER DELIVER.* **2005**, *20*, 978-983.

[43] Martin, D. E.; Wong, W.K.; Dickmander, D.L.; Lee, R.L.; Melvold, D.J. Increasing WSCC power system performance with modulation controls

on the Intermountain Power Project HVDC system. *IEEE T POWER DELIVER*. **1992**, *7*, 1634-1642.

[44] Bahrman, M. P.; Johnson, B. K. The ABCs of HVDC transmission technologies. *IEEE POWER ENERGY M*. **2007**, *5*, 32-44.

[45] Karawita, C.; Annakkage, U. D.; Multi-Infeed HVDC Interaction Studies Using Small-Signal Stability Assessment. *IEEE T POWER DELIVER*. **2009**, *24*, 910-918.

[46] Lie, X.; Vassilios, G. A. VSC Transmission System Using Flying Capacitor Multilevel Converters and Hybrid PWM Control. *IEEE T POWER DELIVER*. **2007**, *22*, 693-702.

[47] Xu, L.; Yao, L. DC voltage control and power dispatch of a multi-terminal HVDC system for integrating large offshore wind farms. *IET RENEW POWER GEN*. **2011**, *5*, 223-233.

[48] Pou, J.; Zaragoza, J.; Rodriguez, P.; Ceballos, S.; Sala, V.M.; Burgos, R.P.; Boroyevich, D. Fast-Processing Modulation Strategy for the Neutral-Point-Clamped Converter With Total Elimination of Low-Frequency Voltage Oscillations in the Neutral Point. *IEEE T IND ELECTRON*. **2007**, *54*, 2288-2294.

[49] Busquets-Monge, S.; Somavilla, S.; Bordonau, J.; Boroyevich, D. Capacitor Voltage Balance for the Neutral-Point- Clamped Converter using the Virtual Space Vector Concept With Optimized Spectral Performance. *IEEE T POWER ELECTR*. **2007**, *22*, 1128-1135.

[50] Zaragoza, J.; Pou, J.; Ceballos, S.; Robles, E.; Ibaez, P.; Villate, J.L. A Comprehensive Study of a Hybrid Modulation Technique for the Neutral-Point-Clamped Converter. *IEEE T IND ELECTRON*. **2009**, *56*, 294-304.

[51] Recio, S.; Pou, J.; Zaragoza, J.; Robles, E.; Villate, J.L.; Martin, J.L. Fault-Tolerant Neutral-Point-Clamped Converter Solutions Based on Including a Fourth Resonant Leg. *IEEE T IND ELECTRON*. **2011**, *58*, 2293-2303.

[52] Saeedifard, M.; Nikkhajoei, H.; Iravani, R. A Space Vector Modulated STATCOM Based on a Three-Level Neutral Point Clamped Converter. *IEEE T POWER DELIVER*. **2007**, *22*, 1029-1039.

[53] Fazel, S. S.; Bernet, S.; Krug, D.; Jalili, K. Design and Comparison of 4-kV Neutral-Point-Clamped, Flying-Capacitor, and Series-Connected H-Bridge Multilevel Converters. *IEEE T IND APPL*. **2007**, *43*, 1032-1040.

[54] Adam, G. P.; Anaya-Lara, O.; Burt, G.M.; Telford, D.; Williams, B.W.; McDonald, J.R. Modular multilevel inverter: Pulse width modulation

and capacitor balancing technique. *IET POWER ELECTRON*. **2010**, *3*, 702-715.

[55] Saeedifard, M.; Iravani, R. Dynamic Performance of a Modular Multilevel Back-to-Back HVDC System. *IEEE T POWER DELIVER*. **2010**, *25*, 2903-2912.

[56] Hagiwara, M.; Maeda, R.; Akagi, H. Control and Analysis of the Modular Multilevel Cascade Converter Based on Double-Star Chopper-Cells (MMCC-DSCC). *IEEE T POWER ELECTR*. **2011**, *26*, 1649-1658.

[57] Marquardt, R. Modular Multilevel Converter: *An universal concept for HVDC-Networks and extended DC-Bus-applications*. In *Power Electronics Conference (IPEC)*, 2010 *International*, Sapporo, June 21-24, 2010, 502-507.

[58] Kontos, E. Control and Protection of VSC-based Multi-terminal DC Networks. M.S. Thesis, Delft University of Technology, Delft, the Netherlands, September 2013.

[59] Van Hertem, D.; Ghandhari, M.; Curis, J. B.; Despouys, O.; Andrée, M. Protection requirements for a multi-terminal meshed DC grid. In *The electric power system of the future: Integrating supergrids and microgrids* CIGRE International Symposium, Bologna, Italy, September 13-15, 2011

[60] Häfner, J.; Jacobson, B. Proactive Hybrid HVDC Breakers: A key innovation for reliable HVDC grids. In *The electric power system of the future: Integrating supergrids and microgrids* CIGRE International Symposium, Bologna, Italy, September 13-15, 2011

[61] Kulkarni, S.; Santoso, S. Interrupting Short-Circuit Direct Current Using an AC Circuit Breaker in Series with a Reactor. *Advances in Power Electronics*. [Online] **2012.** http://dx.doi.org/10.1155/2012/805958 (accessed March 19, 2015).

[62] EURELECTRIC. *Ancillary Services: Unbundling Electricity Products: an Emerging Market*. 2004. http://www.eurelectric.org/Download/Download.aspx?DocumentID=13934 (accessed March 19, 2015).

[63] He, L.; Liu, C.-C.; Pitto, A.; Cirio, D. Distance Protection of AC Grid With HVDC-Connected Offshore Wind Generators. *IEEE T POWER DELIVER*. **2014**, *29*, 493-501.

[64] Hossain, J.; Mahmud, A., Eds. *Large Scale Renewable Power Generation: Advances in Technologies for Generation, Transmission and Storage*; Springer Science & Business Media, 2014.

[65] Hansen, D.; Michalke, G. Fault ride-through capability of DFIG wind turbines, *RENEWABLE ENERGY*. **2007**, *32*, 1594-1610.

[66] Meegahapola, L. G.; Littler, T.; Flynn, D. Decoupled-DFIG Fault Ride-Through Strategy for Enhanced Stability Performance During Grid Faults. *IEEE T SUSTAINABLE ENERGY*. **2010**, *1*, 152-162.

[67] Morren, J.; de Haan, S. W. H. Ridethrough of wind turbines with doubly-fed induction generator during a voltage dip. *IEEE T ENERGY CONVER*. **2005**, *20*, 435-441.

[68] Ramtharan, G.; Arulampalam, A.; Ekanayake, J.B.; Hughes, F.M.; Jenkins, N. Fault ride through of fully rated converter wind turbines with AC and DC transmission. *IET RENEW POWER GEN*. **2009**, *3*, 426-438.

[69] Feltes, A.; Wrede, H.; Koch, F.W.; Erlich, I. Enhanced fault ride-through method for wind farms connected to the grid through VSC-based HVDC transmission. *IEEE T POWER SYST*. **2009**, *24*, 1537-1546.

In: Offshore Wind
Editor: Chester Mendoza

ISBN: 978-1-63482-364-7

Chapter 2

GROUTED CONNECTIONS, HURRICANES, AND THE EVOLVING U.S. OFFSHORE WIND INDUSTRY: ARE CUTTING-EDGE SCIENTIFIC AND ENGINEERING TECHNOLOGIES SUFFICIENT TO MEET POTENTIALLY UNREALISTIC STANDARDS?

Kimberly E. Diamond
Lowenstein Sandler LLP, in the firm's New York, NY office and American Bar Association's Renewable, Alternative, and Distributed Energy Resources Committee in the Section of Environment, Energy and Resources, US

ABSTRACT

North Sea-bordering countries lead the world in offshore wind farm development, with scientific and engineering breakthroughs, as well as pitfalls, being revealed through these countries' offshore wind turbines. The majority of UK and other North Sea offshore wind turbines use monopole foundations. Grouted connections have been used as industry standard for these turbines' foundation designs. The recent UK case *MT Højgaard a/s v. E.ON Climate and Renewables UK Robin Rigg East Limited*, et al. relates to failed grouted connections due to flawed offshore wind turbine foundation design technology in one of the UK's largest wind farms, the Robin Rigg offshore wind farm. Ruling on this case, the UK's High Court of Justice determined that following industry standards for scientific and engineering design is insufficient to shield designers/contractors from liability when a design fails to meet a contractually-mandated "fitness for purpose" requirement for each turbine to achieve a minimum 20 year service life.

While *MT Højgaard* may help to inform the development of the U.S.'s infant offshore wind industry, strict adherence to certain aspects of this international precedent may hinder such industry's effective launch. Most U.S. Wind Energy Areas for offshore wind development are located along the Mid-Atlantic East Coast. Based on recent events and scientific data, hurricanes now travel further up the East Coast than they have historically. Currently, offshore wind turbine blades are being laboratory-tested to improve their designs and increase their resiliency to hurricanes and extreme weather conditions. Nevertheless, such blades, while at the forefront of innovative engineering and scientific technologies, will remain untested in practice due to the U.S. currently having no offshore wind farms. Accordingly, such blades may fail to withstand actual, successive hurricanes over time. If such blades are contractually obligated to meet a minimum 20 year service life "fitness for purpose" performance standard, they may fail to do so.

This chapter analyzes grouting issues impacting North Sea offshore wind turbines with monopole foundations. It also discusses *MT Højgaard*, using it as a springboard for examining contractual obligations when weighed against industry standards. Additionally, it examines scientific data regarding U.S. hurricanes and explores why cutting-edge scientific know-how, combined with state-of-the art engineering technology, may be insufficient to meet legal standards that are aspirational and potentially unrealistic. Finally, it discusses why the ruling and certain facts in *MT Højgaard* may hinder U.S. offshore wind development if the lessons learned from such case are not adopted.

INTRODUCTION

While the recent United Kingdom case *MT Højgaard a/s v. E.ON Climate and Renewables UK Robin Rigg East Limited,* et al. ("*MT Højgaard*") specifically relates to failed grouted connections due to flawed offshore wind turbine foundation design technology in one of the United Kingdom's largest wind farms, the Robin Rigg offshore wind farm, the precedent this case sets and the implications of its specific facts and findings have international implications that may be applied to the U.S. offshore wind industry. As background, North Sea-bordering countries lead the world in offshore wind farm development. As a result, many offshore wind-related scientific and engineering breakthroughs, as well as pitfalls, are revealed through these countries' offshore wind farms' and offshore wind turbines' performances. The majority of the United Kingdom's offshore wind turbines, as well as most other North Sea offshore wind turbines, use monopole foundations. Grouted connections have been used as the industry standard for these turbines' foundation designs. The United Kingdom's High Court of Justice, in its ruling in *MT Højgaard*, determined that following the industry standard for scientific and engineering designs involved in offshore wind turbine foundation construction is insufficient to shield designers/contractors from liability when such designs fail to meet a contractually-mandated "fitness for purpose" requirement that each turbine achieve a minimum 20 year service life. The length of a "fitness for purpose" standard, therefore, can be a critical element in an offshore wind turbine contract, particularly with respect to risks involved for both developer and designer/contractor alike.

While the facts and circumstances of *MT Højgaard* may help to inform the development of the U.S.'s infant offshore wind industry, strict adherence to certain aspects of this international precedent may hinder such industry's effective launch. Most U.S. Wind Energy Areas for offshore wind development are located along the Mid-Atlantic East Coast. Based on recent events and scientific data, hurricanes now travel further up the East Coast than they have historically. Currently, offshore wind turbine blades are being laboratory-tested at special, dedicated, state-of-the-art blade testing facilities to improve their designs and increase their resiliency to hurricanes and other extreme weather conditions. Nevertheless, such blades, while at the forefront of innovative engineering and scientific technologies, will remain untested in practice due to the U.S. currently having no offshore wind farms. Accordingly, such blades may fail to withstand actual, successive hurricanes over time. If such blades are contractually obligated to meet a minimum 20 year service life "fitness for purpose" performance standard, they may fail to do so. As a result, contractual provisions allowing for the future implementation of engineering and scientific advances in blade technologies over incremental periods during the course of the turbine's service life are advisable. The U.S. offshore wind industry's adoption of such a recommended practice will provide for a staggered approach to blade replacement, will allow for flexibility with respect to future engineering and design innovations, and will enable adherence to the contracting parties' obligations. It will also encourage precautionary measures to be built into contracts between developers and designers/contractors so that certain operations and maintenance ("*O&M*") risks, in addition to technology risks, are reduced over the course of the offshore wind turbine's service life. Based on lessons learned from *MT Højgaard*, innovations in contractual provisions, as well as in scientific designs and standards, will enable the U.S. offshore wind industry to preemptively mitigate certain risks and launch itself more effectively.

This chapter analyzes grouting issues impacting North Sea offshore wind turbines with monopole foundations. It also discusses *MT Højgaard*, using it as a springboard for examining contractual obligations when weighed against industry standards. Additionally, it examines scientific data regarding U.S. hurricanes and explores why cutting-edge scientific know-how, combined with state-of-the art engineering technology, may be insufficient to meet legal standards that are aspirational and potentially unrealistic. Finally, it discusses why the ruling and certain facts in *MT Højgaard* may hinder U.S. offshore wind development if the lessons learned from such case are not adopted.

BACKGROUND ON GROUTED CONNECTIONS IN OFFSHORE WIND TURBINES

To understand the significance grouted connections play in an offshore wind turbine, it is important to understand both the physical construct of a monopile wind turbine, as well as the scientific and engineering concepts relating to why grouted connections play a major role in the construction and proper functioning of such turbines. To take advantage of stronger winds at higher altitudes, offshore wind turbines are generally larger than onshore wind turbines, with larger capacities.[1] Presently, European nations lead the world in the installation of offshore wind turbines. As of July 2013, all fully-constructed and operational utility-scale offshore wind turbines were located in Europe, with 65% being located in the North Sea, 16% being located in the Baltic Sea, and 19% being located in the Atlantic Ocean.[2] In particular, the waters bordering the United Kingdom are populated with the vast majority of the world's offshore wind farms. This is due to the United Kingdom's goal to meet 25% of its power needs with renewable energy sources by 2020.

[1] Generally, today's in-service, utility-scale onshore wind turbines range up to 3 megawatts (MW) in capacity, whereas now, due to market innovation and improved designs, offshore wind turbines now are available in 5MW capacities and greater. See The Crown Estate, "A Guide to an Offshore Wind Farm."

[2] See European Wind Energy Association ("*EWEA*") [July 2013].

Offshore wind turbines are scheduled to generate approximately 15% of this energy. The United Kingdom has already been working to satisfy this goal for years. The Crown Estate is the owner and landlord of virtually all the seabed surrounding the United Kingdom, from the shore out to the 12 nautical mile (22.2 kilometers) territorial limit. The Energy Act 2004 granted The Crown Estate vested rights to license wind farm projects on the continental shelf located within the Renewable Energy Zone, out to 200 nautical miles for the purpose of renewable energy generation. The Crown Estate launched the leasing of this seabed for offshore wind farm projects over the course of approximately a decade, beginning in late 2000, through its Round 1, Round 2, Round "2.5," and Round 3 of seabed lease awards.[3]

A number of foundation[4] types have been used for the offshore wind turbines located in these various European offshore wind farms.[5] Notably, the

---

[3] Specifically, The Crown Estate launched the leasing of this seabed for offshore wind farm projects through Round 1 (launched in late 2000; 18 leases awarded to the following wind farms; (i) Blyth; (ii) North Hoyle; (iii) Scroby Sands; (iv) Kentish Flats; (v) Barrow; (vi) Burbo Bank; (vii) Beatrice; (viii) Lynn; (ix) Inner Dowsing; (x) Rhyl Flats; (xi) Gunfleet Sands I; (xii) Robin Rigg (sites A and B); (xiii) Teeside; (xiv) Ormonde; (xv) Tunes Plateau; (xvi) Cromer); (xvii); and Scarweather Sands; (xviii) Shell Flats), Round 2 (launched in 2003; 15 leases awarded to the following wind farms: (i) Gunfleet Sands II; (ii) Docking Shoal; (iii) Race Bank; (iv) Sheringham Shoal; (v) Humber Gateway; (vi) Triton Knoll; (vii) Lincs; (viii) Westermost Rough; (ix) Dudgeon; (x) Greater Gabbard; (xi) London Array (I and II); (xii) Thanet; (xiii) Walney (I and II); (xiv) Gwynt y Mor; and (xv) West Duddon), and Round 3 (launched in 2010; 9 leases awarded to the following wind farms: (i) Moray Firth; (ii) Firth of Forth; (iii) Dogger Bank; (iv) Hornsea; (v) Norfolk Bank; (vi) Hastings; (vii) Isle of Wight; (viii) Atlantic Array; and (ix) Irish Sea). There was also a "Round 2.5" in 2009 offered to Round 1 and Round 2 developers for purposes of allowing them the opportunity to tender for site area extensions to their existing projects (or planned projects, as the case may be), so that the construction on their extension areas could be underway prior to development started on Round 3 projects, thereby allowing for a more stable flow of wind projects in the supply chain prior to Round 3. As a result, the United Kingdom possesses approximately 1,075 currently functioning offshore wind turbines, making it the largest producer of offshore wind energy globally. Patel [Dec. 2014]; GL Garrad Hassan [June 2010]; The Crown Estate. "A Guide to an Offshore Wind Farm"; UKERC [Sept. 2010].

[4] In an offshore wind turbine, the foundation is part of the turbine's support structure which transfers loads acting on the transition piece and pile or piles (as applicable) into the seabed. Collectively, the transition piece and the pile(s) are known as the turbine's "sub-structure." AWEA [Sept. 16, 2012].

[5] For instance, in addition to monopile foundations, tripod and tripile substructures using three short legs connected to one central column, are also used. Denmark's Alpha Ventus wind farm and Bard Offshore I wind farm each use this type of foundation. See Cesari [May 2013]. Jacket substructures, wherein steel frames in a lattice design connected by diagonal or horizontal tubes are drilled into to the seabed using pile foundations, have also been used in a number of offshore wind farms, such as Denmark's EnBW Baltic 2, the United

majority of these turbines each have a cylindrical, monopile foundation, or, rather, a foundation consisting of one single, vertical pile. Monopile turbines are used in shallow water depths, wherein the water depth is 30 meters deep or less, and generally are thought to be the most cost-effective foundation option for turbines up to 5 megawatts (MW). Picture a monopile as a massive, hollow steel tube that is 4.5 meters to 9 meters in diameter that is drilled up to 40 meters into the seabed in water depths of up to 30 meters.[6] A second tube, called the transition piece, supports and connects the offshore wind turbine's tower[7] with the monopile. To accomplish this feat, the transition piece, which has a slightly larger diameter than the monopile, fits over and around the monopile. This area of overlap between the monopile and the transition piece occurs underwater, at the top of the monopile and at the bottom of the transition piece. Together, the monopile and the transition piece constitute the sub-structure of an offshore wind turbine.[8] Grout filling, a fast-curing, high-strength concrete, is the bonding material used to prevent the transition piece from moving.[9] The grout filling is pumped into the annulus between the

---

Kingdom's Ormonde, and Belgium's Thornton Bank II and Thornton Bank III. Cesari [May 2013]. Gravity-based foundation structures are also used for offshore wind turbines. According to the European Wind Energy Association ("*EWEA*"), at the end of 2012, for all offshore wind farms that were on-line and operational, the offshore wind turbines had the following types of foundations (percentages correspond to the proportional share of offshore wind turbines in operation): (i) monopile – 1,376 turbines (74%); (ii) gravity based - 302 turbines (16%); (iii) jacket – 91 turbines (5%); (iv) tripile – 47 turbines (3%); (v) tripod – 29 turbines (2%); and (vi) floating or other – 10 turbines (0%). EWEA [July 2013].

6 See BOEM "Offshore Wind Energy – Commercial Offshore Wind Energy Generation"; Deign [March 18, 2011]. Water constitutes the zone between the water level and the sea floor. The seabed is located beneath the sea floor and is the area under the sea floor in which piles are inserted. How deep a pile is driven into the seabed, or, rather, the distance from the bottom of the pile in its installed position to the sea floor constitutes the "pile penetration." See AWEA [Sept. 16, 2012].

7 The tower, which connects the sub-structure to the rotor-nacelle assembly, houses certain facility components, allows sheltered access for personnel, and is the part of the wind turbine that is connected to the nacelle, an enclosed hollow shell-like piece that houses the wind turbine's shaft, gearbox, and generator. The rotor, consisting of the turbine blades and blade hub) are connected through a drive train to the generator. The turbine's foundation, monopile or otherwise, supports the tower. *See* BOEM "Offshore Wind Energy – Offshore Wind Energy Technology"; MMS, "Technology White Paper" [May 2006]; AWEA [Sept. 16, 2012].

8 AWEA [Sept. 16, 2012].

9 The use of grouted connections in offshore wind farms is a relatively new phenomenon, despite such technique being used since the 1970s for other offshore structures. In 2002, Horns Rev I, located off the coast of Denmark, was the first offshore wind farm to use grouted connections. See *MT Højgaard* at paragraph 10.

monopile and the transition piece to fill the ring-shaped gap between the outer circumference of the monopile and the inner circumference of the transition piece. This connection between the transition piece and the monopile is called the grouted connection, which works by friction and not by adhesion between the steel and the grout.

Monopile offshore wind turbines are subject to severe dynamic loads, making the fatigue capacity of grouted connections very important. The dynamic, axial (or vertical) loads from the wind that impact the turbine's tower are transferred to the grouted connection as a moment produced by a force acting through the length of the transition piece. The moment is transferred as a force-couple into the grouted connection, which compresses (i) one side of the grout at the top of the monopile at the point where the monopile ends but is surrounded by the transition piece and (ii) the other side of the grout at the bottom of the transition piece at the point where the monopile extends up and into the transition piece.[10] Accordingly, the forces and bending moments are transferred from the transition piece to the grout and then to the monopile through horizontal contact forces and the resulting friction, where such forces are then transmitted into the seabed, the soil strata below the sea floor.[11] Cyclic horizontal loads from the wind and waves impacting the offshore wind turbine are transferred to the grouted connection where the stresses are distributed, thereby allowing the turbine's foundation to receive the load transfer through the bending moment. The grouted connection, therefore, must have sufficient strength to transmit the forces and moments to the monopile, must have superior fatigue resistance characteristics to withstand the extensive cyclic loading from the wind and waves, and must serve the purpose of providing structural integrity to the offshore wind turbine's foundation. The grouted connection is also tasked with withstanding the vertical weight of the turbine's tower as well as the lateral force of the wind.

The idea behind the use of a transition piece is that it compensates for imperfections in the angle of the monopile. Monopiles driven into the seabed

[10] See Dedić [Summer 2009].

[11] The field of science that addresses the horizontal and vertical load impacts and such loads' distribution is continuum mechanics, an area that analyzes the behavior of a continuous deformable material subjected to external loadings such as forces, moments, gravitational forces, displacements, velocities, and thermal changes. Grout, under this analysis, is considered as a continuum of closely-packed particles. Under this analysis, differential equations relating to infinitesimal differences in quantities between neighboring particles can be used with respect to grout. See Dedić [Summer 2009].

are generally allowed a tolerance of up to 1 degree off vertical. Therefore, if the seabed shifts or if the inclination of the monopile is otherwise subject to faults up to 0.5 degrees, the transition piece can compensate for this fault, enabling the transition piece and the turbine's tower to remain vertical.[12] The grout serves as a means of rectifying this faulty angularity, so that the transition piece and the tower itself remain vertical. With the grouted connection in place, the transition piece in a monopile turbine becomes the part of the turbine's sub-structure that provides a level connection between the tower and the pile.[13]

Shear keys, or circumferential weld beads, are round, bump-like horizontal (or nearly horizontal) protrusions on the outside surface of the monopile (or inner cylinder) and on the inside surface of the transition piece (or outer cylinder) in the region where the transition piece and monopile overlap. While the grout still remains inserted between the transition piece and the monopile, the shear keys are thought to increase sliding resistance between the grout and the steel of the transition piece and monopile, so that no settlement occurs. Notably, over the last 30 years, the oil and gas industry has used cylindrical-shaped grouted connections with shear keys to connect offshore oil and gas structures to their foundation piles.[14]

The original design standard for offshore wind turbines, DNV-OS-J101, set by certification authority Det Norske Veritas (DNV),[15] was based on limited test data available at the time. A guidance note to this standard advised that it may be conservative to assume that the axial load and the bending moment on a monopile turbine do not interact. Another guidance note for this standard stated that "For connections subjected to bending moment and shear force, no parametric design formulae have yet been established. Therefore, detailed investigations must be carried out for such connections."[16] Together, these guidance notes may be interpreted as meaning that while a simplified formula or model may be supplied as general guidance so that the number of variables of the actual system is reduced, such simplification must be addressed and rectified in practice so that the calculations performed for the

---

[12] See *Windpower* [Sept. 30, 2014].

[13] AWEA [Sept. 16, 2012].

[14] See *Ecology, Environment & Conservation* [2013].

[15] On September 12, 2013, Oslo, Norway-based DNV merged with GL to form DNV GL, an organization that is the resulting organization from the legacy organizations DNV, GL, Garrad Hassan, KEMA, and GL Renewables Certification.

[16] See Dedić [Summer 2009].

real, unique offshore wind turbine system are tailored specifically for that particular system. The DNV-OS-J101 standard also allowed the wind farm developer and its designers/contractors flexibility insofar as they had the option to use shear keys to increase the axial strength properties between the grout and the steel of the transition piece and monopile.[17]

In theory, monopile offshore turbines with cylindrical-shaped grouted connections without shear keys run the risk of incurring flaws in their grouted connection, resulting in fatigue cracks that can proliferate into the steel structure of the turbine itself. Based on data available at the time, numerous offshore wind farm developers and designers/contractors viewed flawed grouted connections as low probability, high risk events. Accordingly, these developers and designers/contractors elected to forego the use shear keys, as shear keys added expense to the project and would increase overall project costs.

During the period from approximately 2009 – 2010, it became evident that the failure rate of grouted connections without shear keys was a more probable event than originally anticipated. Two main factors contributed to this higher-than-expected failure rate. The first factor was a technical flaw in the DNV-OS-J101 industry standard. Based on the limited amount of publicly available data, it was thought that DNV's design standard was faulty and largely the reason why there was an industry-wide epidemic of grouted connection failures and "slippage" incidents. In fact, as discussed herein, it was later discovered that indeed the DNV-OS-J101 design standard that had been in use during and prior to this time period had overestimated the axial capacity of grouted connections without shear keys.[18] The second factor contributing to the epidemic of grouted connection failures was the occurrence of more frequent and greater than anticipated extreme weather conditions. Each offshore wind farm is unique, given its particular location, the underwater topography of the benthic region of the body of water in which it is situated,

---

[17] It has since been shown through subsequent scientific studies that grouted connections with shear keys show remarkable residual resistance against axial loading after fatigue loading in bending. See Klose [June 17-22, 2012].

[18] Previous published standards, such as ISO 19902 (2007) which was used in the petroleum and natural gas industries for fixed steel offshore structures, assumed that the bending moment and the axial loading capacity were independent of one another and do not interact. Monopile designs for offshore wind turbines are driven more by bending moments as compared to structures known from the oil & gas industry. See Klose [June 17-22, 2012].

the exposure to storm surges,[19] the strength of underwater currents, and the surrounding waves' magnitude, height, and frequency, as well as other external conditions.[20] A greater number of violent storms, higher wave heights with stronger forces, and more turbulent winds than originally anticipated challenged current technologies and the ability of current state-of-the-art offshore wind turbine designs to withstand the greater stresses exerted by such conditions.

The combination of the underestimated, heighted extreme adverse weather conditions and the miscalculation in the design standard proved to be a catastrophic combination for most European monopile offshore wind turbines. The error in the DNV-OS-J101 design standard was applied to and incorporated in the actual designs of numerous offshore wind farm facilities. This, in conjunction with the unanticipated high number of extreme adverse weather events of higher intensity than predicted, led to the breaking of the bond between the grout and the steel surface between the monopile and the transition piece in numerous monopile offshore turbines. This breakage, in turn, led to sliding, causing the axial capacity to depend mainly on existing surface irregularities of the monopile and transition piece as well as radial stiffness. Reports from various offshore wind farms located throughout the North Sea during such period indicated that the contact forces between the grouted connections and the steel to which they were connected were failing as the turbines were operating, causing slippage of the transition piece.[21]

This "slippage" issue became a major issue impacting numerous monopile offshore wind turbines and offshore wind farms alike, particularly in wind farms around the United Kingdom. It was estimated that approximately four-fifths of Europe's installed offshore wind turbines suffered from dissolved and cracked grout, causing these turbines to shift on their foundations and no

[19] A storm surge is characterized by wind and atmospheric pressure variations causing irregular movements in the impacted body of water. AWEA [Sept. 16, 2012].

[20] These include environmental conditions, such as exposure to storms, their frequency, strength, and associated turbulence, as well as other climatic factors such as snow, ice, and temperature. AWEA [Sept. 16, 2012].

[21] According to *Windkraftjournal* (2010), the first reported case of slippage occurred in June 2009 during routine inspections at the Egmond aan Zee offshore wind farms in The Netherlands. All 36 monopile turbines in the array were inspected, and a number of slippage instances were detected. In 2010, more cases from other European offshore wind farms became public. See Klose [June 17-22, 2012].

longer stand vertical.[22] Remediation measures to fix these grouting failures is substantial, costing approximately €120,000, or approximately $150,000,[23] per turbine for corrective retrofits. While these figures may appear high, they are likely still much lower than the real costs of repair. This is because only limited information with respect to remediation costs, as well as limited information and actual scientific data gathered from the impacted wind farms, has been made public. "Slippage," then, caused the United Kingdom's offshore wind industry as a whole to suffer a real setback during a rapid expansion period for offshore wind farms, particularly since such industry was playing and continues to play a key role in helping the United Kingdom achieve its target renewable energy goals.

It is not unusual for a standard, such as one used in the wind industry, to be updated as a matter of regular procedure with best practices from the entire industry.[24] In the offshore wind industry, failed grouted connections on numerous offshore turbines led DNV to update and impose a modified DNV-OS-J101 industry design standard.[25] In October 2009, DNV initiated a joint industry project (JIP) with other partners so that the result of such project could be used to revise, update, and improve upon the current design standard

---

[22] Storms that are more violent and that are of greater magnitude and frequency than originally anticipated are, in part, responsible for causing North Sea monopile offshore wind turbines to experience bending movement within their foundation parts, causing certain of these turbines to tip. See *Windpower* [Nov. 8, 2013]. In a September 2010 Europe wind energy advisory note, IHS Emerging Energy Research stated that approximately 600 of Europe's 948 installed offshore wind turbines See Deign [2011]. In July 2012, almost two years later, a paper was delivered at the International Offshore and Polar Engineering Conference, indicating that the number of grouted connection failures for grouted connections without shear keys had increased to approximately 800 reported instances. See Klose [July 17-22, 2012]. European offshore wind farms that had been constructed prior to 2009 and had used monopoles as their foundations included the following (name of wind farm (year wind farm became operational and country)): (i) Arklow Bank (2007, Ireland); (ii) Barrow (2005, United Kingdom); (iii) Blyth (2000, United Kingdom); (iv) Egmond aan Zee (2006, Netherlands); (v) Gunfleet Sands (2010, United Kingdom); (vi) Horns Rev 1 and Horns Rev 2 (2002 and 2009; Denmark); (vii) Kentish Flats (2005, United Kingdom); (viii) Lely (1994, Netherlands); (ix) Lynn & Inner Dowsing (2008, United Kingdom); (x) North Hoyle (2003, United Kingdom); (xi) Princess Amalia/Q7 (2008, Netherlands); (xii) Rhyl Flats (299, United Kingdom); (xiii) Robin Rigg (2010, United Kingdom); (xiv) Scroby Sands (2004, United Kingdom); (xv) Utgrunden (2000, Sweden); and (xvi) Walney (2011, United Kingdom). See Cesari [May 2013].

[23] All dollar references in this chapter refer to U.S. dollars.

[24] See Project Finance International [May 6, 2010].

[25] In 2013, HIS Emerging Energy Research reported that grouting had started to dissolve on over 60% of Europe's offshore wind turbines, causing them to shift on their foundations. See *Windpower* [Nov. 8, 2013].

and methodology and to decrease risks for potential damage that could occur as a result of the current standard. In January 2011, DNV led a second joint industry project (JIP) with 11 offshore wind developers to update and improve upon existing knowledge and design practice by investigating the current design standard in effort to establish a more reliable design methodology.[26] This second JIP was conducted because it was thought that the design standard DNV had set previously did not adequately account for the physical behavior of the grouted connections. As a result, this second JIP suggested a new assessment methodology for grouted connections used in monopile offshore wind turbines that were not originally built with shear keys. This JIP suggested a modification to the current design standard, so that the grouted connection section would be conical-shaped to account for large dynamic bending moments on the monopiles. Accordingly, no longer did the industry-wide design standard recommend a purely cylindrical-shaped design for grouted connections without shear keys as additional supports for the axial load. Such JIP continued until its completion in May 2012 and included fatigue testing of grouted test specimens in laboratory settings as well as finite element analyses. The findings and results from such testing and analyses ultimately resulted in DNV's recommendation of a revised design methodology for monopile offshore wind turbines.

THE RISKS OF STANDARDS AND "FITNESS FOR PURPOSE" REQUIREMENTS: *MT HØJGAARD* SETS AN INTERNATIONAL PRECEDENT

On April 15, 2014, the United Kingdom's High Court of Justice, Queen's Bench Division, Technology and Construction Court set an international precedent for the offshore wind industry through its decision in the case *MT Højgaard a/s v. E.ON Climate and Renewables UK Robin Rigg East Limited,* et al. While *MT Højgaard* relates specifically to allocating responsibility for remedying failed grouted connections, it is significant due to the precedent it sets with respect to risks associated with using a commonly accepted

[26] Since 1864, DNV's role as an independent authority has been to assist in solving complex technical challenges that society and business face. DNV plays an important role for the wind industry by developing new methodologies, standards, and practices to enhance the wind industry overall. See Bertnes [2011].

scientific/engineering industry standard and due to the pitfalls associated with agreeing to contract terms that bind the designer/contractor to an obligation to satisfy a "fitness for purpose" requirement.

The background of this case is as follows. In 2006, E.ON Climate Renewables UK Robin Rigg East Ltd and E.ON Climate and Renewables UK Robin Rigg West Ltd (collectively, *"E.ON"*), the defendants and project developer[27] of the Robin Rigg offshore wind farm,[28] hired the claimant, MT Højgaard, a designer/contractor,[29] for the design, fabrication, and installation of the foundations for all 60 monopile offshore wind turbines comprising the Robin Rigg offshore wind farm in England's Solway Firth, a body of water constituting part of the Irish Sea that forms part of the border between England and Scotland and which is located northeast of the Isle of Man, northwest of Cumbria, and south of Dumfries and Galloway. Each turbine in the Robin Rigg wind farm possesses a foundation consisting of a monopile and a transition piece.[30] A grouted connection 6.45 meters in length holds each monopile and transition piece together.[31] The court noted that grouted connections are "a well-established method of connecting tubular supports for offshore structures."[32] Notably, shear keys were not used on any of Robin

[27] In the offshore wind industry, a "developer" is a party or parties that hold the responsibility of permitting, planning, constructing, and commissioning the particular offshore wind farm's turbines. AWEA [Sept. 16, 2012].

[28] The Robin Rigg wind farm is one of the Crown Estate's Round 1 of offshore wind farm projects, the first of successive rounds of wind farm projects that the Crown Estate has hosted to date. There are actually two separate sites for this wind farm – Robin Rigg A and Robin Rigg B. UKERC [Sept. 2010]. Each of the 60 Robin Rigg offshore wind turbines is a 3MW turbine. Accordingly, this wind farm, which is located in an area where the maximum water depth is 12 meters, has a capacity of 180MW (60 turbines x 3MW = 180MW). Each turbine in the Robin Rigg array has a total height of 125 meters, with a rotor diameter of 90 meters. See 4C Offshore "Robin Rigg – Developers/Owners/Operators". The Crown Estate is the owner of virtually all of the seabed surrounding the UK, from the shore out to the 12 nautical mile territorial limit. The Energy Act 2004 granted The Crown Estate vested rights to license wind farm projects on the continental shelf located within the Renewable Energy Zone, out to 200 nautical miles for the purpose of renewable energy generation. The Crown Estate. "A Guide to an Offshore Wind Farm."

[29] In the offshore wind industry, a "designer" consists of one or more parties who are responsible for the design of the offshore wind turbines or its component parts, such as electric service platforms and cables. AWEA [Sept. 16, 2012].

[30] Each monopile in the Robin Rigg array has a diameter of just over 4 meters, each with a slightly larger diameter transition piece that is approximately 8 meters long. See *MT Højgaard* at paragraph 2.

[31] See *MT Højgaard* at paragraph 3.

[32] *See id.*

Rigg's turbine foundations. As set forth in the contracts for the design and construction of the Robin Rigg offshore wind turbines, such turbines, including their foundations, were supposed to have a service life of 20 years. MT Højgaard prepared a design for the Robin Rigg turbines' foundations based upon the 2004 DNV-OS-J101 international standard for the design of offshore wind turbines ("*J101 (2004)*"), the primary standard for offshore wind turbine design at that time. It was later discovered that the value attributed to a variable used in one of the equations in J101 (2004) was underestimated by a factor of approximately 10. At the time that MT Højgaard designed the Robin Rigg turbines' foundations, neither MT Højgaard nor anyone else in the offshore wind industry was aware of this error. MT Højgaard's designs for the Robin Rigg wind farm were implemented between mid-2006 and mid-2007. In 2009, movement was discovered in the grouted connections that had been designed in accordance with the J101 (2004) industry standard. It was at that time that the error in one of the J101 (2004) equations was discovered. As a result of the faulty grouted connections, all 60 Robin Rigg turbines exhibited signs of movement, and all required remedial work. The agreed upon cost of such work was €26,250,000 or $32,739,290. E.ON claimed that MT Højgaard should be held liable for the failed grouted connections due to MT Højgaard's warranting that the turbine foundations would have a 20 year service life and due to MT Højgaard's failure to use reasonable skill and care in applying J101 (2004).

Among the many facts that the court considered, the following course of events and the following language and phrases in the Contract (as defined below) were the ones of greatest significance. In May 2006, MT Højgaard was invited to tender its bid for contract with respect to the foundation work on the Robin Rigg wind farm. It was a condition of the Instructions to Tenders that the tender was to be completed in accordance with the Employer's Requirements of E.ON, in its role as the employer. Winning the tender, MT Højgaard, in its role as the designer/contractor, entered into a contract for services with E.ON in December 2006 (the "*Contract*"). The Contract contained nine parts along with Annexes to the Employer's Requirements. Part C of the Contract contained the List of Definitions. Two of the definitions were of great significance. The first term was "Fit for Purpose," which was defined as meaning "fitness for the purpose in accordance with, and as may properly be inferred from, the Employer's Requirements."[33] The second term was "Good Industry Practice," which was defined as meaning:

> "in relation to any particular undertaking or task . . . those standards, practices, methods and procedures . . . to be performed with the exercise of skill, diligence, prudence, and foresight that can ordinarily be expected from a fully skilled contractor who is engaged in a similar type of undertaking or task in similar circumstances consistent with recognized international standards."[34]

Part D of the Contract contained the Conditions of Contract, which included a Section 8.1, "General Obligations," which stated the following:

> "The Contractor shall, in accordance with this Agreement, design, manufacture, test, deliver and install and complete the Works: (i) with due care and diligence expected of appropriately qualified and experienced designers, engineers and constructors (as the case may be) . . . (iv) in a professional manner . . . in accordance with . . . Good Industry Practice . . . (viii) so that the Works, when completed, comply with the requirements of the Agreement . . . (x) so that each item of Plan and the Works as whole shall be fit for its purpose as determined in accordance with the Specifications using Good Industry Practice . . . (xv) so that the design of the Works and the Works when completed by the Contractor shall be wholly in accordance with this Agreement and shall satisfy any performance specifications or requirements of the Employer as set out in this Agreement . . . .[35]

Parts G, H, I, J, and K of the Contract consisted of the Employer's Requirements. Section 3 of Part I of the Employer's Requirements contained the Technical Requirements for the Design Basis (Wind Turbine Foundations) (the "*Technical Requirements*"). A number of significant provisions appeared in such Section 3. For instance, in the "Introduction" in Section 3.1, the following provisions appeared:

---

[33] *See id.* at paragraph 22.

[34] See *id.*

[35] See *id.* at paragraph 23.

> "It is stressed that the requirements contained in this section and the environmental conditions given are the MINIMUM requirements of the Employer to be taken into account in the design. It shall be the responsibility of the Contractor to identify any areas where the Works need to be designed to any additional or more rigorous requirements or parameters."
>
> "The Contractor shall undertake the detailed engineering design of all structures, fixtures and fittings required to comply with the requirements of this Agreement and the Conditions of Contract. The Contractor shall assume full responsibility for design and installation of the structures." [36]

Also, Section 3.2.6, entitled "Life," stated that "All parts of the Works, except wear parts and consumables shall be designed for a minimum service life of 20 years."[37] Section 3b.5.1, entitled "Scope," also emphasized the importance of a 20 year service life through the following language: "The design of the structures addressed by this Design Basis shall ensure a lifetime of 20 years in every aspect without planned replacement. The choice of structure, materials, corrosion protection system operation and inspection programme shall be made accordingly." Additionally, Section 3b.5.6, entitled "Service Life," also contained a 20 year service life provision, as follows: "All parts of the Works, except wear parts and consumables shall be designed for a minimum service life of 20 years."[38]

The "Technical Requirements" that appeared in Section 10 of Part I of the Contract contained a Section 10.2.5, entitled "Structural Design," and a Section 10.5.1, entitled "Grouted Connection." Each of these sections contained provisions mandating that no scheduled maintenance would be required during the operational life of the wind farm. These sections also allowed the Contractor discretion as to whether to employ shear keys within the grouted connection.[39]

It is well-accepted that as a matter of sound business practice, and so that there is a meeting of the minds of the respective parties to the contract, the specific standard for performance needs to be clearly set forth within the body of the contract. Because the parties to a contact negotiate and agree upon a

---

[36] See *id.* at paragraph 25.

[37] See *id.* at paragraph 27.

[38] See *id.* at paragraph 28.

[39] See *id.* at paragraph 29.

particular standard, such agreed-upon standard is the crucial factor that determines the commerciality of the transaction and the intent of the parties to the contract at the time the contract was made. In fact, it is well-settled under British law that courts, when interpreting a particular provision in a commercial contract, endeavor to determine what the parties to such contract meant through the language such parties selected as well as through the analysis of what a reasonable person would have understood the parties to have meant by such language.[40] For instance, in the British case *Rainy Sky SA v. Kookmin Bank*, Rainy Sky SA and other claimants entered a claim against defendant, Kookmin Bank for $46,620,000 with respect to a shipbuilding contract. At issue was the interpretation of a particular phrase of the contracts that Kookmin Bank had entered into with claimant and the other shipbuilders. In rendering its opinion, the court held that to interpret a contractual provision, one must (i) use considerations of "business common sense" to determine what the parties meant, and (ii) determine what a reasonable person would have understood the parties to have meant by such language, a reasonable person being someone who "has all the background knowledge which would reasonably have been available to the parties in the situation in which they were at the time of the contract."[41] The *Rainy Sky SA* court clarified further that in instances where the terms of a contract are open to more than one interpretation, "it is generally appropriate to adopt the interpretation which is most consistent with business common sense."[42] Moreover, the High Court of Justice Chancery Division in *Pink Floyd Music Limited, Pink Floyd (1987) Limited v. EMI Records Limited*, during its analysis of the parties' intent with respect to certain phrases' interpretation with respect to royalty payments in a Master License Agreement, held that "the true construction" of a clause in the contract must be interpreted in the context of the contract as a whole.[43]

Given this precedent from *Rainy Sky SA*, *Pink Floyd*, and a similar line of cases, despite MT Højgaard's very well-made arguments as to why claimant's reliance on the J101 (2004) industry standard should determine on the outcome of the case, the court held in favor of E.ON, based on the above-quoted language of the Contract between E.ON and MT Højgaard. MT Højgaard's position was that its obligation was to exercise "Good Industry

[40] See *Rainy Sky SA* at 2900.

[41] See *id.* at 2907.

[42] See *id.* at 2911.

[43] See *Pink Floyd* at paragraph 46.

Practice" by producing a design compliant with J101 (2004), without taking into account the possibility that J101 (2004) might contain a fundamental error. The court disagreed with this position. Specifically, the court determined that the "purpose" of the works referenced by the phrase "fit for its purpose" in Section 8.1(x) of the General Obligations was determined by the Technical Requirements for the project. Good Industry Practice in such Section 8.1(x), consequently, referred to the determination of the purposes and did not qualify the obligation as to fitness for purpose. The court noted that it was clear under the provisions of the Technical Requirements that a 20 year service life was the purpose that needed to be achieved under the terms of the Contract. Because the grouted connections in all turbines in the Robin Rigg wind farm had failed within two to three years of such turbines' installation, the court stated that "[p]lainly, if the grouted connections of the foundations had to have a service life of 20 years, they have failed to achieve it."[44] According to the court, per the language of the Contract, MT Højgaard had an obligation to produce a work product that would meet a "fitness for purpose" design standard – a standard that incorporated and went beyond the lesser standard of carrying out the design with reasonable care and skill. As the court noted, because the Technical Requirements of Section 3.1 required a foundation designed to last a "minimum" service life of 20 years, it was necessary for MT Højgaard, as the designer/contractor, to create a work product that would last for period longer than merely the design life period fit for the purpose of complying with J101 (2004). Any design failing to meet the purpose of lasting for a minimum of 20 years would be considered faulty, defective, and a breach of the Contract's provisions. As a result, based on the Contract's terms, the court determined that MT Højgaard had warranted to E.ON that the offshore turbine foundations would have a service life of 20 years, that E.ON was entitled to rely on such warranty, and that MT Højgaard, despite its designing the grouted connections in accordance with the J101 (2004) industry standard so that they were fit for the purpose of complying with such standard, had breached such warranty due to the grouted connections failing within two to three years after the turbines' construction.

[44] See *MT Højgaard* at paragraph 67.

QUESTIONS *MT HØJGAARD* RAISES FROM LEGAL, RISK/REWARD, SCIENTIFIC, AND TECHNOLOGICAL PERSPECTIVES

The offshore wind industry is still a relatively young industry that continues to evolve, with new technological breakthroughs and state-of-the-art scientific discoveries that continue to occur. Advances in engineering and other scientific areas have enabled turbine designs to evolve rapidly, replacing what was previously thought to be the most high-tech, revolutionary products with newer, cutting-edge products. These advances also cause widely accepted industry standards to be subject to modification, based on improved data or new knowledge. From an O&M perspective, it has been acknowledged and accepted that wind turbines generally are designed less conservatively than other more traditional machines, as wind turbines' designs are often closer to the acceptable edge of the design criteria.[45] Accordingly, offshore wind developers and their designers/contractors alike assume a certain amount of technological risk every time they agree to use a particular scientific or engineering standard, as that standard on which agreed-upon work product is predicated is based on the best available knowledge at the time. This is why it is crucial for project developers, designers/contractors, and other parties to a contract to read, understand, and carefully draft such contract's provisions that set forth the contract's ultimate purpose. As the *MT Højgaard* case illustrates, contractual provisions stating the contract's purpose are likely to trump compliance with an industry standard if mere compliance with such standard falls short of fulfilling that contract's purpose.

The *MT Højgaard* case is also interesting as it raises a host of law, science, and technology questions. First, it raises the issue of whether certain contractual terms, while agreed-upon between the contracting parties, are aspirational and present technological impossibilities, rather than being realistically based on the current state of technology. For example, the J101 (2004) industry design standard for offshore wind turbines consisted of engineering and scientific formulas that were more theoretical than field-tested at the time the standard was used. The formula in J101 (2004) was the most cutting-edge formula for its intended purpose, based on the best available data and scientific knowledge at the time. Such formula was only found to be

[45] See Alewine [Dec. 2014].

defective after its practical application in the field. The scientific community cannot know that a standard is faulty in practice unless such standard is actually field-tested in authentic settings. Accordingly, the minimum 20 year service life requirement for grouted connections may not have been a realistic contractual provision; rather, it may have been one that was aspirational.

This leads to the second issue of when a court should step in and void a contractual provision for impossibility of purpose, and when parties to a contract should be bound to contractual provisions that only have a remote likelihood of being fulfilled. It is generally not a court's place to know when the threshold level for impossibility of purpose has been reached, particularly with respect to scientific formulas that require theorizing about their proper application or about whether they were accurately derived. This is why the parties to a contract, rather than the court, are in the best position to gauge what is an impossibility versus what is possible, though unlikely, particularly with respect to the application of scientific data in unique locations. In *MT Højgaard*, a 20 year turbine life possibly may have been attainable if shear keys had been used, and if MT Højgaard, as designer/contractor, had double-checked and re-calculated the defective J101 (2004) formula. Under *MT Højgaard*, then, while not advisable, parties to a contract involving the application of scientific data have the right to mutually agree to ambitious contract terms that have a remote probability of being accomplished.

This brings us to the third and most tricky issue that *MT Højgaard* raises: the issue of how much risk each party to a contract should be willing to undertake, from both the perspective of an offshore wind developer and from the perspective of a designer/contractor. To structure a wind farm project so that it is profitable and provides a reasonable return to the wind farm's investors, the financial model for the wind farm project has to be carefully calculated and researched. From an economic perspective, keeping costs low where possible but still meeting the necessary project specifications is the most efficient manner in which to proceed to meet these objectives.[46] Balancing design and component reliability with ways to maximize project profitability, then, are elements a developer must consider. In *MT Højgaard*, E.ON, as developer, was willing to take a calculated risk of not mandating that shear keys be incorporated as part of the Robin Rigg turbines' foundation design. The industry standard did not require taking this precautionary measure. As a result, E.ON did not require that such precaution be taken,

[46] See *id.*

particularly due to the additional expense that taking such precautions would have entailed. Although unknown from the court record, E.ON could have been of the mind that foregoing shear keys would not result in the occurrence of a Black Swan event – an event that is extremely difficult to predict and has a very low probability of occurring, carries a tremendously high risk of resulting in damage of a very large magnitude, and uses hindsight to make its occurrence not appear to be a surprise.[47] In retrospect, had E.ON known that foregoing shear keys could lead to the Black Swan event of all of 60 of the Robin Rigg turbines tilting on their foundations and requiring multi-millions of dollars to remedy, then E.ON would likely have taken the precaution of mandating that shear keys be used in the turbines' foundation design. One take-away from *MT Højgaard* is that offshore wind developers need to weigh and give careful consideration to which precautionary measure to take. A small investment in a particular precautionary measure may result in saving an exponentially greater amount of money in the future, due to the repercussions and ramifications associated with foregoing such measure being avoided.

Moreover, *MT Højgaard* illustrates the conundrums a designer/contractor faces with respect to winning and losing a competitive bid process for a contract with an offshore wind developer, as the designer/contractor's willingness to undertake a greater amount of risk may mean the difference between winning and losing the bid process. It is unknown from the record if MT Højgaard would have priced itself out of the competitive bid process if it had included shear keys in its proposed turbine foundation design. *MT Højgaard*, then, indirectly demonstrates the dangerous balance a designer/contractor must strike when it makes a bid. Specifically, the designer/contractor must weigh the pros and cons of the following two approaches. First, there is the conservative approach that entails the designer/contractor incorporating precautionary measures, which may mean submitting a higher overall bid price to account for the additional cost of

[47] For a Black Swan event to occur, (i) the event must be a very big surprise that is unprecedented and unexpected at the point in time that it occurred, (ii) the event must have a major effect and deviate beyond what is typically expected to occur in a given situation, and (iii) in hindsight, an explanation for the occurrence can be provided that would have otherwise made such event explainable, predictable, and not surprising. *See* Black Swan and Black Swan Theory. It is thought that the phrase "Black Swan" originated from the Western belief that all swans were white because only white swans had ever been seen. In 1697, though, Willem de Vlamingh, a Dutch explorer, discovered black swans in Australia. This was an unexpected scientific event that significantly changed zoology. See Black Swan: About – What are Black Swan Events?

equipment, parts, and services needed for such precautionary measures to be implemented. Second, there is the more risky approach of making a bid that foregoes a more robust – and likely more expensive – design that incorporate precautionary measures. Because this more risky approach does not need to allot for additional equipment, parts, and services, a lower bid price can be submitted to win a contract with the developer. Taking this latter approach means the designer/contractor assumes the risk that certain adverse events will not occur as a result of the precautionary measure's omission. The risk/reward analysis in which both the developer and the designer/contractor engage, consequently, can have a significant impact on the overall viability and integrity of the offshore wind farm at issue.

IMPLICATIONS OF *MT HØJGAARD* FOR U.S. OFFSHORE WIND FARM DEVELOPMENT

It is important to bear in mind the contextual background behind *MT Højgaard* relative to the current state of the U.S. offshore wind industry, as such context is instructive for purposes of U.S. offshore wind farm construction. First, the Robin Rigg offshore wind farm is located in the United Kingdom, one of the primary markets for offshore wind development.[48] According to a report by The Crown Estate, the owner of the seabed surrounding the United Kingdom and the grantor of leases with respect to wind farm construction, "There is no single way to build and operate an offshore wind farm, and indeed the challenges of scale, water depth and distance from the shore are such that the optimal solutions are still being developed."[49] Also, as The Crown Estate has indicated, in today's market, an average offshore wind turbine may cost approximately £6,000,000, or approximately $9,800,000, for a single 5MW turbine.[50] Second, there are currently no U.S. offshore wind farms. Given the cost of a single offshore wind turbine, it will be advantageous to the U.S. offshore wind industry to take and encourage measures that will make offshore wind turbines as resilient as

---

[48] Globally, the primary markets for offshore wind farm development are as follows: (i) Denmark; (ii) Germany; (iii) The Netherlands; (iv) Sweden; and (v) the United Kingdom. GL Garrad Hassan [June 2010].

[49] See The Crown Estate, "A Guide to an Offshore Wind Farm."

[50] See *id.*

possible based on lessons learned from the United Kingdom's offshore wind farms.

Given that the U.S. offshore wind industry is still in its infant stage, from a lessons learned perspective, the international precedent that *MT Højgaard* sets and the legal, scientific, and risk-based questions that it raises may serve as a source of guidance for U.S. offshore wind developers, designers/ contactors, and courts alike. For developers and designers/contractors, *MT Højgaard* highlights the importance of agreeing on contractual provisions that are realistic, rather than aspirational, in nature. *MT Højgaard* also illustrates that if there is a contractual dispute, a court may hold the contracting parties to terms of the contract on which they agreed, particularly those terms setting forth the contractual purpose or articulating the "fitness for purpose" requirement that must be fulfilled. As discussed above, even if such purpose is unrealistic or aspirational, and even if complying with current industry standards is insufficient to satisfy such contractual purpose, a court may nevertheless hold such parties to the terms of their agreement. These takeaways offer practical guidance for offshore wind developers and designers/contractors to consider before entering into contracts relating to offshore wind turbine construction.

For those offshore wind developers and designers/contractors aspiring to build offshore wind farms in the U.S., tempering contractual provisions so that they are realistic based on current state-of-the–art technologies, rather than so they are purely aspirational in nature, may be the most practical approach and one founded on best practices. This is particularly the case with respect to two things: first, satisfying "fitness for purpose" requirements relating to equipment design, such as for offshore wind turbine blades, and second, to establishing the realistic performance life period for how long such equipment should last. The Contract in *MT Højgaard* correctly articulated the accepted industry standard that an offshore wind turbine generally should have a minimum operational service life expectancy of 20 years.[51] It is logical, then,

[51] According to the U.S. Department of Interior's Minerals Management Service, an offshore wind turbine is expected to have an operational life of 20 – 25 years. MMS [May 2006]. Most onshore wind turbines are designed to meet or exceed that expectation based on the wind loads as defined by IEC 61400-1 (design requirements to ensure that wind turbines are engineered appropriately against damage from hazards within the turbine's planned lifetime). Alewine [Dec. 2014]. Today, offshore wind turbines are designed in accordance with the international offshore wind turbine standard IEC 61400-3. Quarton [2005]. For a wind turbine to achieve a 20 year life expectancy, having a tower and foundation designed in accordance with good engineering practices is critical. Alewine [Dec. 2014].

that offshore wind turbines in the U.S. also would be expected to have a minimum operational service life of 20 years. Accordingly, it stands to reason that U.S. offshore wind developers would want blades for such turbines to fulfill a minimum 20 year operational service life requirement and would likely want such a requirement incorporated into a contract between itself and a designer/contractor or blade manufacturer.

In the U.S., however, imposing a 20 year operational service life for offshore wind turbine blades may be difficult, if not impossible, to achieve, given the unique extreme weather conditions, such as hurricanes, to which many U.S. offshore wind turbines may be subject and to which offshore wind turbines in the North Sea and Europe have never been exposed. This is why incorporating into contracts requirements for much shorter blade service life than 20 years may be the most optimal way to proceed. Such provisions will allow for scientific data on hurricanes to be more appropriately addressed and for breakthrough technologies and advances in blade designs to be used, despite their costs.

THE CURRENT STATE OF THE U.S. OFFSHORE WIND INDUSTRY: WIND ENERGY AREAS, HURRICANE RISKS, AND ADVANCES IN BLADE TECHNOLOGY AND DEVELOPMENTS

While there are no operational commercial-scale offshore wind farms in the U.S., federally-approved Wind Energy Areas, or "WEAs," have been designated off the U.S.'s Atlantic East Coast as zones for offshore wind project development, with the U.S. Department of Interior's Bureau of Ocean Energy Management (BOEM) having issued seven commercial wind energy leases on the Outer Continental Shelf (OCS) in such areas to date.[52] Following developer input, stakeholder feedback, and input from other sources, in

[52] In 2009, President Barack Obama announced the final regulations for the OCS Renewable Energy Program, which was authorized by the Energy Policy Act of 2005. Such regulations provide a framework for offshore wind lease issuances. BOEM is responsible for offshore wind development in federal waters. To date, the commercial wind energy lease areas for which leases have been granted are as follows: (i) Cape Wind Lease Area; (ii) RI/MA North Lease Area; (iii) RI/MA South Lease Area; (iv) Delaware Lease Area; (v) MD – North Lease Area; (vi) MD – South Lease Area; and (vii) Virginia Lease Areas. *See* Barkoff [Feb. 2011]; BOEM [Nov. 2014].

February 2011, U.S. Secretary of the Interior Ken Salazar and U.S. Secretary of Energy Steven Chu announced the creation of four Mid-Atlantic WEAs on the Outer Continental Shelf off the coasts of Virginia, Maryland, Delaware, and New Jersey, respectively.[53] The creation of these WEAs was viewed as a key component of the "Smart from the Start" initiative for advancing the installation of utility-scale offshore wind turbines in U.S.[54] According to the American Wind Energy Association's (AWEA's) U.S. Wind Industry Annual Market Report Year Ending 2013, there are currently eight WEAs, with WEAs off the coasts of Massachusetts, Rhode Island, and North Carolina, respectively, joining the East Coast states off of which the original WEAs are located.[55] Collectively, these WEAs are poised to allow wind farms built inside of them the ability to take advantage of the approximately 900,000MW of wind energy resource potential that is located off the New England Coast and Mid-Atlantic Coast.[56] Within these WEAs, there are six offshore wind projects currently being planned, with other offshore wind projects being planned outside these WEAs in Atlantic coastal areas off the shores of New York and Maine.[57]

As a basis for comparison, WEAs are similar to the United Kingdom's respective numbered zones in each of The Crown Estate's Rounds for offshore wind farm development.[58] For each of these Rounds, in addition to considerations of shipping density and other navigational risks,[59] developers'

---

[53] See Barkoff [Feb. 2011]; BOEM [Nov. 2014].

[54] See *id.*

[55] See AWEA Market Report [2013].

[56] See MMS [May 2006].

[57] These contemplated projects are as follows (state – project name): (i) Maine – Aqua Ventus; (ii) Massachusetts – Cape Wind; (iii) Rhode Island – Block Island Wind Farm and Deepwater Wind Energy Center; (iv) New York – LI-NYC Offshore Wind Project; (v) New Jersey – Atlantic City Wind Farm; (vi) Delaware – Mid-Atlantic Wind Park; and (ii) Virginia – Virginia Offshore Wind Technology Acceleration Project. Notably, the U.S.'s first one-eighth scale model, 65-foot offshore wind turbine, named "VolturnUS," was deployed in June 2013 off the coast of Maine by the University of Maine's DeepCwind Consortium. *See* AWEA Market Report [2013].

[58] For example, in The Crown Estate's Round 3, the following zone numbers correspond to the related offshore wind farm: (i) Zone 1 – Moray Firth; (ii) Zone 2 – Firth of Forth; (iii) Zone 3 – Dogger Bank; (iv) Zone 4 – Hornsea; (v) Zone 5 – East Anglia; (vi) Zone 6 - Southern Array; (vii) Zone 7 – West Isle of Wight; (viii) Zone 8 – Atlantic Array; and (ix) Zone 9 – Irish Sea. Diamond, et al. [Oct. 2011].

[59] Shipping density refers to the estimated number of vessels per day that pass through a certain area of interest. Variations in shipping density are often shown through a shipping density

and other key stakeholders' input was crucial in the revision and establishment of such zones. For example, following developer input and input from other key government stakeholders during The Crown Estate's Round 3, it was realized that one southern zone lay over one of the United Kingdom's major shipping lanes.[60] This input had significant benefit. In 2008, the Round 3 zone locations were changed, three zones were eliminated, and a new zone was added, creating Round 3's nine zones that exist today.[61]

From the United Kingdom's example, the U.S. learned that shipping and navigational risks need to be addressed early and strategically when offshore wind project siting is being determined. As a result, input from key stakeholders and others has shaped the location and size of WEAs. For instance, certain WEAs were scaled back from their originally proposed locations and sizes, as they were located very near established vessel traffic routes used in foreign trade.[62] A good example of this was when Massachusetts Secretary of Energy and Environmental Affairs, Richard K. Sullivan, requested the U.S. Department of Interior's Bureau of Ocean Energy Management, Regulation, and Enforcement (BOEMRE)[63] to cut the WEA located south of Martha's Vineyard to less than half its original size to accommodate shipping lanes and established fishing grounds. BOEMRE complied, basing its response on scientific evidence and feedback from industry stakeholders.[64]

In addition to WEA's respective locations being determined as a result of input from developers, stakeholders, and others, WEAs' locations are significant due to the weather conditions to which they are subject. All WEAs are located within 50 nautical miles of shore, in immense areas that the U.S.'s National Renewable Energy Laboratory (NREL) identified as having a wind power classification of "Excellent" or "Outstanding," and where wind speeds

model, wherein a grid of cells are color-coded to indicate the number of vessels per day that pass through each such cell. See *id.*; Anatec [Aug. 2, 2002].

60 See *id.*

61 See *id.*

62 Due to vessels being forced to deviate from traditional courses to go around an offshore wind farm, an area that previously did not have high vessel traffic may find itself having increased vessel traffic. Increased vessel traffic on well-traveled vessel routs may increase the probability for a higher incidence of vessel collisions. See *id.*

63 BOEMRE was formerly known as the Minerals Management Service (MMS). On October 1, 2011, BOEMRE's duties with respect to the management of the U.S.'s resource on the outer continental shelf were allocated to the Bureau of Ocean Energy Management (BOEM). See Bureau of Ocean Energy Management, Regulation and Enforcement.

64 See *id.*

are among the highest in the country.[65] While NREL has completed a preliminary offshore study from wind resource assessments of coastal states in five key geographic regions, the focus of this article will be on one particular region: the East Coast.[66] The East Coast region consists of three subparts: (i) New England, (ii) the Mid-Atlantic, and (iii) the South Atlantic Bight. The New England sub-region consists of areas of the Atlantic Ocean located off the shores of Maine, New Hampshire, Massachusetts, Rhode Island, Connecticut, and New York – an area in which three WEAs are located. Throughout this sub-region, winds are generally "Outstanding," with a wind power class of Class 6.[67] This sub-region is generally characterized by lower intensity hurricanes.[68] This sub-region is also prone to nor'easters, storms that travel along the East Coast, generally occur in winter, and are cyclones that have hurricane-force winds accompanied by heavy snow and rain.[69] The Mid-Atlantic sub-region consists of the areas off the coasts of New Jersey, Delaware, Maryland, Virginia, and North Carolina – an area in which five WEAs are located – and, according to NREL, is exposed to winds that are generally "Excellent" or "Outstanding," with wind power classes of Class 5 and Class 6.[70] Similar to the New England sub-region, the Mid-Atlantic sub-region is also prone to nor'easters. The South Atlantic Bight sub-region consists of the Atlantic Ocean offshore areas of South Carolina, Georgia, and Florida.[71] Wind speeds in this sub-region are lower than those in the New England and Mid-Atlantic sub-regions. The downfall, then, with Atlantic Coast WEAs situated in the Mid-Atlantic and New England sub-regions is that both sub-regions are subject to hurricanes, otherwise known as tropical

---

65 According to an NREL map of the U.S.'s national offshore resource, these locations at a 50 meter height boast wind speeds of 16.8 – 17.9 mph in the "Excellent" wind power class (Class 5), and boast wind speeds of 17.9 – 19.7 mph in the "Outstanding" wind power class (Class 6). Musial, et al. [Sept. 2010].

66 These five key geographic regions are as follows: (i) the East Coast (including New England, the Mid-Atlantic states, and the South Atlantic Bight); (ii) the West Coast (California and the Pacific Northwest); (iii) the Great Lakes region; (iv) the Gulf of Mexico; and (v) Hawaii. Musial, et al. [Sept. 2010].

67 See Musial, et al. [Sept. 2010].

68 See *id.*

69 See Diamond [Fall 2012].

70 See Musial, et al. [Sept. 2010].

71 See *id.*

cyclones.[72] While hurricanes are more usual and are of great concern in the South Atlantic Bight, no WEAs are currently contemplated in this sub-region. [73] However, moderate hurricanes do, indeed, frequent the Mid-Atlantic sub-region where WEAs are situated in the greatest number.[74]

Precautionary measures must be taken for wind farms to be sited inside such Mid-Atlantic WEAs, due to such wind farms' increased risk for hurricane exposure. The examination of scientific data with respect to the frequency with which hurricanes make their way up the Atlantic Coast, as well as background about the unique, extreme weather conditions inherent to hurricanes, is therefore instructive for why taking precautionary design measures is necessary for offshore wind turbines that will be built in these areas. The World Meteorological Organization (WMO) has noted that 2001 – 2010 was the warmest decade in recorded history.[75] During this decade, in terms of precipitation, large parts of the Northern Hemisphere, particularly the eastern United States, experienced wetter-than-average conditions.[76] Separately, the National Center for Atmospheric Research (NCAR) has noted that aerosols and sulfates from fossil fuel combustion increase atmospheric acidity while causing a cooling atmospheric effect.[77] NCAR has also noted that carbon dioxide in the atmosphere causes a warming effect.[78] During the 1960's and 1970's, the atmosphere's carbon dioxide content used to "balance out" the cooling effect of the aerosols and sulfates that were also in the atmosphere.[79] Since the 1970's, due to laws regulating aerosol and sulfate emissions, the amount of these chemicals in the atmosphere has been reduced.[80] What this means, though, is that there is no longer a cooling effect to balance out the warming effect from the carbon dioxide in the atmosphere.[81] The surface of the world's oceans absorbs this additional atmospheric heat, causing global

---

[72] A tropical cyclone is a closed atmospheric or oceanic circulation around a zone of low pressure that originates over tropical ocean regions. A tropical cyclone gets classified as a tropical depression, tropical storm, or hurricane, based on its wind speed. AWEA [Sept. 16, 2012].

[73] See Musial, et al. [Sept. 2010].

[74] See *id.*

[75] World Meteorological Organization [March 23, 2012].

[76] See *id.*

[77] See Holland [Aug. 11, 2014].

[78] See *id.*

[79] See *id.*

[80] See *id.*

[81] See *id.*

oceanic temperatures to rise.[82] Because hurricanes are fueled by warm, moist air, a rise in air and ocean temperatures higher up the U.S.'s East Coast than has been the case historically could result in hurricanes traveling even further up the East Coast and lasting longer in the future.[83] Wind farms in WEAs, consequently, will be at increased risk for being hit by at least one hurricane over the course of their 20 year operational lives, including a "Black Swan" hurricane that could cause massive devastation.

With respect to wind speeds and hurricanes' historical paths up the U.S. East Coast, recent scientific data also provides cause for concern for wind farms located in Mid-Atlantic WEAs. Hurricanes are characterized by extreme winds that can impose high instantaneous loads, multi-directional winds, winds of sustained, extreme wind speeds, high wind/wave frequency, rapid changes in wind direction, and impulsive gust loading.[84] Hurricanes are rated on the National Oceanic and Atmospheric Administration's (NOAA's) National Hurricane Center's Saffir-Simpson Hurricane Wind Scale – a scale that assigns hurricanes a Category 1 – 5 rating based on each hurricane's sustained wind speed intensity.[85] For instance, Category 1 hurricanes have winds that range from 74 – 95 mph, are considered very dangerous and will cause some damage.[86] Category 2 hurricanes have winds that range from 96 – 110 mph, are considered extremely dangerous, and will cause extensive damage. Hurricanes that are Category 3, Category 4, and Category 5 hurricanes, whose winds range from 111 – 129 mph, 130 – 156 mph, and 157 or higher mph respectively, are considered major hurricanes, as they will cause devastating or catastrophic damage.[87] In fact, with respect to hurricanes, NCAR has observed that since the 1970's, there has been a substantial increase every decade thereafter in the frequency and intensity of hurricanes; specifically, in every part of the world except for the East Pacific Basin, the world is experiencing two and a half times as many Category 4 and Category 5 hurricanes than it did approximately 30 years ago, particularly with respect to Category 4 hurricanes.[88] Atlantic hurricanes are tropical cyclones that form in

[82] See Avery [Aug. 11, 2014].

[83] See Diamond [Fall 2012].

[84] See Musial, et al. [Sept. 2010].

[85] See National Hurricane Center, "Saffir-Simpson Hurricane Wind Scale"; Diamond [Fall 2012].

[86] See National Hurricane Center, "Saffir-Simpson Hurricane Wind Scale".

[87] See *id.*

[88] See Holland [Aug. 11, 2014].

the Atlantic Ocean, Gulf of Mexico, and Caribbean Sea, areas collectively constituting the North Atlantic Basin.[89] The 2001 - 2010 decade was also characterized by the highest level of tropical cyclone activity on record for the North Atlantic Basin.[90]

Recent history and scientific data also bear out the fact that hurricanes do, in fact, hit areas in the Mid-Atlantic and New England regions, on the mid- to upper East Coast, particularly in the Mid-Atlantic region. Given trends in atmospheric and oceanic warming, hurricanes may continue to hit such areas with increased frequency in future years. The Atlantic hurricane season runs from June 1 – November 30 annually.[91] Based on the National Weather Service's historical data, the average path of a hurricane traveling up the Atlantic Coast and hitting WEAs during these months is "likely" to occur in June and July, with chances of hurricanes hitting these areas increasing to "more likely" from August through October.[92] For example, in August 2011, Hurricane Irene, a major hurricane, struck the eastern seaboard from North Carolina up to New York with its pounding winds, causing extensive devastation to impacted areas.[93] Hurricane Irene traveled through the Mid-Atlantic WEAs during its journey. Just one year later, in October 2012, Hurricane Sandy, another major hurricane, traveled through the Mid-Atlantic WEAs before making landfall on the New Jersey coast. Based on historical data for hurricane paths, as well as these relatively recent historical events, it is likely, if not probable, that one or more hurricanes will hit offshore wind turbines in WEAs located off the Atlantic Coast, particularly those situated in Mid-Atlantic WEAs. The ability of wind turbines in these WEAs to not only withstand hurricanes but to continue to perform well following one or more hurricane strikes will depend substantially on the resiliency of the designs incorporated into such turbines and the products from which such turbines are constructed.

While there has been some evidence that wind farms located near the Atlantic shoreline can survive hurricane-force winds, hurricanes operate differently when they are over water than they do when they are over land. For instance, the five 1.5MW turbines in the Jersey Atlantic Wind Farm, located at

---

[89] See Diamond [Fall 2012].

[90] See World Meteorological Organization [March 23, 2012].

[91] See National Hurricane Center. "Tropical Cyclone Climatology".

[92] See *id.*

[93] See National Hurricane Center. "Past Track Seasonal Maps."

the ACUA Wastewater Treatment Plant in Atlantic City, New Jersey, was subjected to Hurricane Sandy.[94] Authorities from Infigen Energy, the operators of these turbines, stated that the turbines were put into "hurricane mode," wherein they were taken off line and rotated so that they faced into the wind to allow for the least resistance to Sandy's winds.[95] All five turbines emerged from Sandy with no physical damage.[96] However, it was noted that although the turbines remained in "good shape," there nevertheless were "minor issues" of an unspecified nature that needed to be worked out.[97] Moreover, the dynamics of a hurricane are different when they are solely above water than when they make landfall. Hurricanes are fueled by the heat of the water and water evaporating from the water's surface.[98] When hurricanes make landfall, they are no longer being fueled from such power sources, causing their winds gusts to slow down. So, while the Jersey Atlantic Wind Farm was subjected to winds of approximately 64 mph (sub-hurricane-status winds) from Sandy,[99] such onshore wind farm did not encounter the full brunt of Sandy's hurricane-force winds experienced over the Atlantic Ocean. Additionally, offshore wind farm turbines' blades are generally much longer than onshore wind turbines' blades, making their length, weight, and differences in materials from which they are made factors that distinguish them from onshore turbines' blades for purposes of post-hurricane survivability and functionality.

Unfortunately, current technology may not allow for survivability and sustained performance for offshore wind turbines located off the Mid-Atlantic Coast, particularly with respect to their blades. AWEA's Offshore Compliance Recommended Practices (2012) for the design, deployment and operation of offshore wind turbines in U.S. waters mandates compliance with the International Electrotechnical Commission's IEC 61400-3 code for offshore turbines, as supplemented by other guidance documents (including its parent document, IEC 61400-1), recommended practices, and/or standards.[100] A

[94] See ACUA. "Jersey-Atlantic Wind Farm Fact Sheet."

[95] See ACUA. "Atlantic City Windfarm Withstands Force of Hurricane Sandy."

[96] See Brownen [Jan. 20, 2014]; ACUA. "Atlantic City Windfarm Withstands Force of Hurricane Sandy."

[97] See ACUA. "Atlantic City Windfarm Withstands Force of Hurricane Sandy."

[98] See Diamond [Fall 2012].

[99] See Brownen [Jan. 20, 2014]; ACUA. "Atlantic City Windfarm Withstands Force of Hurricane Sandy." Because Hurricane Sandy was effectively not a hurricane when it made landfall on New Jersey's coast, it was named "Superstorm Sandy" to distinguish its strength and magnitude from its prior hurricane status.

[100] See AWEA [Sept. 16, 2012].

Carnegie Mellon University research team found that offshore wind turbines placed in U.S. waters may be vulnerable to hurricane-force extreme winds because offshore wind turbines are currently only designed to withstand wind speeds characteristic of Category 1 hurricanes.[101] Also, according to an NREL report, "the current generation of wind turbines that are type-certified to the International Electrotechnical Commission's IEC 61400-3 code for offshore turbines may not be able to withstand the extreme loading caused by hurricanes."[102] This is because while the IEC standards require the designer/ contractor to determine the extreme loads the hurricane conditions will impose and to design the turbines with the appropriate structural strength, most wind turbines are designed and type-certified well in advance of the exact site location for the turbine.[103] In Atlantic Ocean waters prone to hurricanes, the extreme wind conditions will potentially exceed the extreme conditions specified in IEC 61400-1 Class 1A – the most stringent class for turbine design – which does not specifically address hurricanes and assumes that wind gusts will not exceed 70 m/s, or approximately 157 mph.[104] It is projected that Category 4 and Category 5 hurricanes, including "50-year storms," will have gusts greater than 80 m/s, or approximately 179 mph.[105] Currently, wind industry executives and engineers maintain that a Class 1 turbine, designed with current technology, should be able to withstand a Category 3 hurricane.[106] However, whether an actual offshore wind turbine can handle the loads from a Category 3 hurricane, and what level of incremental damage turbine blades and gears will sustain after being subjected to repeated exposure to such extreme weather conditions, remain unknown. This means that current offshore wind turbines on the market today need design modifications to enable them to be more hurricane-resistant and to improve their survivability.

Engineers and scientists will need to improve blade design so that offshore wind turbine blades, as part of the overall offshore turbine design, will incorporate technological advances at the cutting edge of design technology and be able to withstand and function well following encounters with hurricanes. Because an offshore wind turbine is larger than an onshore wind

---

[101] See Swaney [Feb. 14, 2012].

[102] See Musial, et al. [Sept. 2010].

[103] See *id.*

[104] See *id.;* AWEA [Sept. 16, 2012]. The next edition of IEC 61400-3, on which work has already begun, will incorporate explicit guidance for hurricanes.

[105] Musial, et al. [Sept. 2010].

[106] Diamond [Fall 2012].

turbine, its blades extend further than do those of an onshore wind turbine.[107] For instance, blades for a 5MW offshore wind turbine are generally over 60 meters long and over 5 meters wide at their broadest point, weighing between 15 – 25 tons.[108] Due to their size, materials, and design features, offshore wind turbine blades are quite expensive. A single turbine blade for a 5MW offshore wind turbine can cost roughly £250,000 – £350,000, or approximately $390,000 – $550,000.[109] Blade bearings, which enable the adjustment of the blade pitch angle to control the turbine's power output, minimize loads, and stop or start the turbine as applicable, are also costly, with a single blade bearing for a 5MW turbine costing roughly £40,000 – £50,000, or approximately $62,000 – $78,000.[110] The pitch system itself, which actually adjusts a blade's pitch angle, costs roughly £100,000 – £150,000, or $157,000 – $235,000 for a 5MW turbine.[111] Offshore wind turbine blades typically are designed with certain factors in mind. For instance, different regions of these blades are designed to handle both fatigue and extreme loading, along with cyclically varying aerodynamic and reversing gravity loading.[112] Extreme loading for these turbines can come from the high number of hours that the turbine operates in a turbulent wind field, or from storm loading.[113] Offshore wind turbine blades are also designed with a lightning protection system, which allows lightning that strikes these blades to pass safely to the turbine's nacelle and tower and therefore provides a certain level of protection for the blades and the rest of the turbine.[114] For purposes of assessing blade damage following hurricanes and other severe storms, physical inspection of the blades focused on damage to the blades and blade degradation is necessary.[115]

---

[107] An average utility-scale onshore wind turbine has a tower height of approximately 60 – 80 meters, with blades that are approximately 30 – 40 meters long. In contrast, an average commercial offshore wind turbine has a tower height of above 61 meters and blades that area approximately 76 to 107 meters long. MMS [May 2006].

[108] See The Crown Estate, "A Guide to an Offshore Wind Farm."

[109] See LikeForex.com. "Currency Converter Calculator."

[110] See *id.*

[111] See The Crown Estate, "A Guide to an Offshore Wind Farm." LikeForex.com. "Currency Converter Calculator."

[112] See The Crown Estate, "A Guide to an Offshore Wind Farm."

[113] See *id.*

[114] Different offshore wind turbine blade suppliers employ different strategies for the capture and transfer of high voltage currents from the impact point to the hub and balance of the turbine, based on the materials from which the blades are made and other aspects of blade design. See The Crown Estate, "A Guide to an Offshore Wind Farm."

[115] See AWEA [Sept. 16, 2012].

In contrast to the Black Swan grouting issue that arose in North Sea monopile turbines, the issue of blade survivability in hurricanes has already been identified, and the U.S. government has employed research and development as a strategy to address the destructive forces that hurricanes present with respect to blade resiliency and to improve blade performance, all before actual field usage of such blades. As illustration, in 2009, the U.S. Department of Energy (DOE) granted $24,700,000[116] to provide the State of Massachusetts with the balance of the $40,000,000 financing needed to build the Massachusetts Clean Energy Center (MassCEC), a first-of-its-kind large-blade testing facility that is 50,000 square feet and capable of testing blades up to 80 meters, created in partnership with NREL.[117] Today, the MassCEC Wind Technology Testing Center (WTTC) in Charlestown, Massachusetts, located at the Boston Autoport in Boston Harbor,[118] is the largest of three blade-testing facilities in the U.S.,[119] providing the full gamut of turbine blade testing for blades up to 90 meters, or approximately 300 feet, in length – almost the size of an American football field – which will be used on 5MW turbines,[120] to ensure that such blades meet the international certification requirements for international design standards, including meeting the IEC 61400-23 standard for static and fatigue tests[121] and passing lightning protection testing.[122] Having the WTTC facility means that blades longer than 50 meters can now be tested in the U.S., rather than in Europe, which, according to U.S. Secretary of Energy Steven Chu, will help ensure that the most efficient wing turbines

---

116 These federal funds were granted under the American Recovery and Reinvestment Act. The balance of funding for MassCEC come from a $2,000,000 federal NREL grant, and $13,200,000 in grants and loans from the State of Massachusetts's Clean Energy Center. Graff [July 2010]; O'Brien [May 2011].

117 See Musial, et al. [Sept. 2010]; MassCEC. "Wind Technology Testing Center;" Graff [July 2010].

118 See Patrick [May 2009].

119 The other two blade testing facilities are in Maine and Colorado. The Charleston facility, though, is the largest of its kind. Oaks, et al. [May 2013]

120 As a basis for comparison, 1.5MW and 2MW turbines generally have blades that are 200 feet in length. Graff [July 2010].

121 Static testing determines the blade's strength, so that the blade's breaking point can be determined. Approximately 190 sensors can be attached to a blade for such testing purposes, to determine stresses, strains, and how the material from which the blade is made is behaving. Such testing takes approximately two weeks to complete. Fatigue testing simulates the wear and tear of a blade over what would be the equivalent of a 20 year period, and can take from three to six months to complete testing. Graff [July 2010]; Oaks, et al. [May 2013]; O'Brien [May 2011].

122 See MassCEC. "Wind Technology Testing Center."

are built in the U.S.[123] Moreover, at WTTC, engineers also are engaged in developing the latest wind turbine prototype methodologies to improve reliability, evaluate the most advanced technologies in advanced composite and other materials used in blade design, and actualize the deployment of "next generation" offshore wind turbine technologies.[124] At the beginning stages of the WTTC, WTTC's Executive Director at the time, Rahul Yarala, stated that, "With [WTTC], we are not only catching up [with the other countries that have large-scale testing centers], but we are going to get ahead of the curve."[125]

While WTTC is dedicated to accelerating the success of clean energy technologies,[126] WTTC also is tasked with the unenviable job of improving blade design to minimize the gap between blade productivity on one hand and blade reliability, resiliency, and functionality on the other. This means that engineers at WTTC will need to innovate blade designs that enable offshore wind turbine blades to be hurricane-resistant, yet still able to function in a manner that allows them to produce an optimal amount of energy on a regular basis. Having a blade design that will survive hurricanes, but will not allow for the generation of an optimal – or at least, expected – quantity of energy is not a desirable outcome. Likewise, a blade design that produces an optimal amount of energy but whose structural integrity cannot withstand repeated exposure to hurricanes is also undesirable. Reaching a middle ground between these two extremes may be challenging. From a research and development perspective, WTTC serves an important function insofar as its findings and improved understanding of the extreme loads from hurricanes can help inform blade design so that offshore wind turbine blades can be built in a manner such that these blades will not only survive a hurricane, but will remain operational and fully functional thereafter as well as after bouts with other hurricanes that follow. Moreover, the price of innovation for maximizing energy production while improving blade life cycle survivability could be very expensive, as such technological improvements may mean added expenses in terms of the cost of advanced materials, the blade size and shape, the number of blades used on an individual turbine, etc. For instance, an innovation in overall offshore turbine design could include having turbines that possess only two blades that are

[123] See Patrick [May 2009].

[124] See MassCEC. "Wind Technology Testing Center;" Graff [July 2010].

[125] See Graff [July 2010].

[126] See O'Brien [May 2011].

longer than those on the market today and rotate downwind of the turbine's tower. In fact, such turbine and blade designs are being contemplated today with respect to 10MW turbines; it is contemplated that such turbines would have a rotor diameter of approximately 150 meters, using only two blades designed from advanced blade materials.[127] At the present time, however, blades tested at WTTC, while at the forefront of engineering and scientific design, only will have been laboratory-tested and will remain untested in practice in the U.S. Consequently, such blades may fail to withstand actual, successive hurricanes over time, thereby disallowing them from meeting a "fitness for purpose" performance standard if their respective offshore wind turbines must meet a contractual requirement of a minimum 20 year service life. Consequently, the risk of blade failure due to extreme weather conditions may outweigh any monetary or other benefit for offshore wind blade manufacturers or designers/contractors.

LESSONS LEARNED FROM *MT HØJGAARD*: HOW CUTTING-EDGE TECHNOLGIES CAN BE USED TO MITIGATE "FITNESS FOR PURPOSE" REQUIREMENTS

In terms of blade design and technologies, compliance with industry standards, and agreement to a "fitness for purpose" contractual performance standard, the facts and events of *MT Højgaard* are instructive and can inform U.S. offshore wind development from a lessons learned perspective. As the facts of *MT Højgaard* illustrate, offshore wind turbine designs walk a thin line that separates acceptable, cutting-edge design technologies from defective products of flawed designs. Offshore wind turbine designs push the acceptable thresholds of scientific and engineering design, transforming what was previously thought to be technological impossibilities into technological breakthroughs. In the offshore wind industry, many developers and designers/contractors in practice do not have the luxury of having their products field tested. Rather, they must rely on theoretical calculations and smaller scale test models that are not generally exposed to the unique, on-site conditions experienced at a particular wind farm site. Accordingly, certain

---

[127] For instance, materials such as carbon fiber and titanium are contemplated for use in these blades, as such materials allow for a higher strength to mass ratio, which assists in handling increased loading. UKERC [Sept. 2010].

international standards that dictate how offshore wind turbines and certain of their parts must be designed may be based on the best available scientific data. Nevertheless, due such standards' lack of actual field testing, such standards may be either flawed or aspirational in nature, thereby making such standards unrealistic in practice.

However, *MT Højgaard* also stands for the proposition that offshore wind developers and designers/contractors do not necessarily need to take tremendous technological risks every time they agree to use a particular scientific or engineering standard in offshore wind turbine construction. Rather, steps can be taken to mitigate such risk. For example, in *MT Højgaard*, the J101 (2004) standard employed the use of scientific formulas that were theoretical in nature and not field-tested, based on the best available data and scientific knowledge at the time. This standard, while itself partially flawed, recommended the use of shear keys as an optional precautionary measure. E.ON could have taken this more expensive precautionary measure of mandating the use of shear keys in its turbine designs. E.ON took a calculated risk, gambling that the probability of damage caused by foregoing this precautionary measure was low enough to warrant not incorporating shear keys into its turbine designs. Had E.ON known about the frequency and magnitude of the actual adverse weather conditions to which its turbines would be subject, it could have taken such precautionary measure, and could have avoided the Black Swan event that plagued the Robin Rigg turbines.

In contrast to the situation in *MT Højgaard*, the U.S. offshore wind industry, including wind farm developers and designers/contractors, have the advantage of possessing knowledge that can guide their behavior and decisions with respect to the election of precautionary measures taken with respect to turbine design for offshore wind turbines that will be located in WEAs. While IEC 61400-3 and IEC 61400-1 may not currently incorporate specifications or mandate certain safeguards for turbine blades exposed to hurricanes, sufficient data exists and is publicly available with respect to historical hurricane paths and magnitude. This, coupled with information available about how rising atmospheric and ocean surface temperatures can impact future hurricane frequency and intensity, can provide developers and designers/contractors alike with sufficient knowledge about risks involved and precautionary measures to take for wind farms situated in WEAs. In particular, while there may not be a specific industry standard or guidance note currently in place today that mandates certain requirements for offshore wind turbine blades that will likely encounter hurricanes, developers and designers/contractors are now in a position to give careful consideration to which precautionary measures to take in terms of blade design. Developers have the advantage of knowing that engineers at WTTC are discovering ways to improve blades so that new designs can be invented that are at the forefront of engineering and scientific technology. They also are in a position to weigh which precautionary measures to take with respect to hurricanes, particularly in terms of blade design selection, as certain blades at this juncture may be more hurricane-resilient than other blades, but may also be less efficient for energy production purposes. Moreover, the price of innovation can be costly, impacting blade prices and, consequently, the overall cost of each turbine as well as the overall costs of the wind farm's construction. Developers need to make the difficult strategic decision of whether the risk of one or more actual hurricane encounters is outweighed by the price of these state-of-the-art blades, bearing in mind that a small investment in a precautionary measure up front may save a significantly greater amount of damage repair costs in the future.

MT Højgaard is also instructive for guidance purposes with respect to the election and agreement upon certain contractual provisions, particularly those involving a "fitness for purpose" performance standard. From a designer/contractor's perspective, *MT Højgaard* highlights the importance of agreeing on realistic contractual provisions – ones that have a very high likelihood of being satisfied and that are based on current technology and scientific ingenuity. With respect to blade designs for offshore wind turbines in waters off the U.S. East Coast, a blade designer/contractor who agrees to a "fitness for purpose" performance standard such that the blades are required to last and perform properly for the minimum 20 year operational service life of the turbine is taking on a significant risk as fulfillment of such contractual provision may be difficult, if not impossible, to achieve.

One possible approach for addressing a "fitness for purpose" provision with respect to blade design is to build a temporal element into the contract, such that at certain incremental periods over the course of the turbine's operational life, a blade can be subject to replacement or upgrade, with no penalty for such replacement being attributed to the designer/contractor. Drawing on the concept of "re-powering" from the onshore wind industry,[128]

---

[128] Re-powering in the onshore wind industry means removing the original turbines on a specific location and replacing them with newer turbines that incorporate the latest technologies, which are generally larger than the original turbines. The removal of the original turbines and the installation of the new turbines can be done in several phases. Also, a re-powering project generally means that more megawatt hours (mw/h) of power can be produced on the same property with fewer turbines. See DOI Right-of-Way (R/W) Grant for Mesa Wind Power Corporation [Feb. 13, 2008].

wherein state-of-the-art advances in turbine design and blade technologies are used to replace outdated and relatively obsolete turbines and blades, a contract having such a provision would allow for successive removal and replacement phases, wherein the original blades are removed and replaced with improved, next generation successor blades, and so forth. In contrast to a re-powering, though, the removal and replacement phases would occur during, rather than at the end of, the turbine's operational life. For example, having five removal and replacement phases would enable an offshore wind turbine to be outfitted with its original blades for a period of four years, until its immediate successor blades are installed. A similar process would occur over the course of the turbine's 20 year operational life, thereby enabling a turbine to take advantage of next generation blade designs that incorporate improved, state-of-the-art technologies every four years. This incremental blade replacement process will enable the developer and wind farm to take advantage of innovations that make blades more hurricane-resilient and/or able to produce more power than the original blades during adverse weather conditions. Moreover, such a forward-looking provision would enable the turbines and the wind farm in which they are located to evolve at a pace that tracks more closely the pace of evolving technologies in the industry. Each such upgraded turbine would be an improvement upon an offshore turbine that is confined to use during its entire 20 year operational life the original blade technology that was only available at the beginning of such life.

Incorporating temporal, incremental "phased blade replacement" requirements into contracts will also help improve a designer/contractor's risk profile by quantifying and minimizing the duration of risk a designer/ contractor may face if such risk is tied to a particular set of blades. This means that if a liability mechanism is tied to the original blades' performance, the designer/contractor's risk exposure will be confined only to the relatively short-term, fixed period during which the blades are in service. This, in turn, reduces the length of risk exposure for such original blades from a 20 year risk to a four year risk. The same four year risk period would apply to each successive phase of new blade replacements/installations. This type of risk profile presents a more favorable scenario from an O&M perspective for the designer/contractor particularly toward the end of the turbine's 20 year life cycle. As illustration, in year 16, under the "phased blade replacement" scenario, new blades would be installed on the turbines, so that the designer/contractor is relying on the performance of replacement blades that incorporate 16 years of design improvements and technological advancements in extreme weather resiliency, rather than relying on the performance of the turbine's original blades installed 16 years earlier that likely will be more prone to exhibiting signs of wear and tear, needing repair, or experiencing complete failure. A modification of this "phased blade replacement" scenario could include removing and replacing only half or one-third of the wind farm's turbines blades over a fixed period of time, and having the balance of blades replaced at some fixed period thereafter, so that the blade replacement periods are staggered and potentially make the overall blade replacement process less costly. Additionally, building in the "phased blade replacement" concept into a contract in a manner that is not punitive in nature may enable the developer and the designer/contractor to foster a better, less adversarial working relationship over the life of the wind farm's operations. In the wind industry today, the cost of a visit for O&M purposes to one or more turbines can be quite costly. While the cost of blade replacement may potentially rival or exceed that of regular and unplanned O&M visits over the course of an offshore wind turbine's lifetime, a willingness to adopt and negotiate provisions for the potentially more expensive "phased blade replacement" approach up front may help to make such visits less punitive and more amicable in nature, particularly if a fair, negotiated rate or fee cap and floor for such visits are included as contract terms.

Also, unlike the situation of the Robin Rigg wind farm wherein all 60 of such wind farm's turbines are tilting on their foundations and require multi-millions of dollars to remedy, designers/contractors – as well as developers – are in a position to avoid having all offshore turbines in a particular U.S. offshore wind farm from being tied to a dated industry standard that could be flawed, and to take advantage of improvements to this standard based on identified miscalculations or other flaws that have been revealed over the course of time through actual field usage of offshore wind turbine blades. Because blade replacement can be costly, and because it is possible that not all turbines in a particular wind farm would suffer blade defects in need of simultaneous repair, this type of incremental blade-replacement scenario may also be palatable to the wind farm developer. Accordingly, incorporating "phase blade replacement" requirements for blade replacements into contracts where there is a "fitness for purpose" standard with respect to a turbine's 20 year operational service life may be an optimal way to proceed. Despite the up-front cost of blade replacements, having such contractual provisions would allow for the allocation of risk over a much shorter blade performance life, improve performance-based O&M risk, and potentially mitigate against more costly blade replacements that may be needed if such approach is not taken.

CONCLUSION

Case law decisions from countries that have mature, robust offshore wind industries can be instructive from not only a purely legal perspective but for purposes of informing behavior and offering insight with respect to certain aspects of offshore wind industry development. Rulings relating to the offshore wind industry from the United Kingdom – which has one of the world's most developed offshore wind industries – can be used as guidance from a lessons learned perspective for countries such as the U.S. looking to launch its offshore wind industry. The *MT Højgaard* case from the United Kingdom is no exception. The holding in *MT Højgaard* makes clear that agreed-upon terms between contracting parties, such as a "fitness for purpose" requirement, is sufficient to override generally accepted industry standards. If a 20 year service life is contractually mandated for all parts of offshore wind turbines, including turbine blades, strict adherence to such requirement may hinder the U.S. offshore wind development process. This is because based on current scientific and engineering technology, offshore wind turbine blades are not yet resilient enough to enable them to survive, sustain minimal damage, and perform reliably after one or more hurricane strikes. For offshore turbines that will be erected in wind farms located in WEAs along the U.S.'s East Coast, this is of particular concern, as scientific data from the last few decades and historical events from recent years indicate that hurricanes are likely to travel further up the East Coast, occur with more frequency, and be of greater magnitude in the future. As *MT Højgaard* illustrates, the most advanced, cutting-edge, and agreed-upon industry-wide engineering and scientific standards – such as the J101 (2004) industry design standard for monopile offshore wind turbine foundation design – may be flawed in some key respect when applied in practice, as testing simulations may not be rigorous enough to accurately parallel real-world extreme weather conditions. Similarly, despite advances being made in blade design at blade testing facilities such as WTTC, such cutting-edge scientific know-how and state-of-the art engineering technologies may be insufficient to meet a U.S.-specific or an international offshore wind industry standard that requires such blades to withstand exposure to certain categories of hurricanes, as the ability to satisfying such standards – which are in the process of being set – may be potentially aspirational and unrealistic at this juncture.

At a minimum, *MT Højgaard* illustrates that contractual terms allowing for some degree of fluidity between current state-of-the-art designs and related standards on one hand and designs and standards of the future that are based on more evolved scientific and engineering technology on the other is a more preferable approach than agreeing on terms that are static, rigid, and do not allow for fluidity and evolution in engineering and scientific designs. While a minimum 20 year operational life may be industry standard for the life of an offshore wind turbine, as illustrated in *MT Højgaard*, mandating a minimum 20 year term to satisfy a "fitness for purpose" standard may not be ideal or advisable for certain components of an East Coast U.S. offshore wind turbine, such as blades, that are at risk for not meeting such a standard. Strict adherence to such a term would lead to several non-optimal options for the U.S. offshore wind industry. The first option is to forego building offshore wind turbines until the state of scientific technology has developed offshore wind turbine blades that, with relative certainty, can survive and perform reliably after one or more hurricane strikes, to avoid failing a 20 year "fitness for purpose" service life requirement in an offshore wind turbine contract. This option may potentially allow for industry standards and certain contractual obligations to be met, but will have the effect of delaying an expeditious launch of the U.S. offshore wind industry until yet-to-be invented hurricane-resilient blade technologies are invented. The second option is to proceed with the launch of the U.S. offshore wind industry, building offshore wind turbines with present state-of-the art blade engineering technology while knowing that the offshore wind turbine blades run a risk of encountering hurricanes over their minimum 20 year operational life. Accordingly, for the U.S. offshore wind industry, particularly with respect to blade design, modifying what may otherwise be typical "fitness for purpose" contractual provisions so that the service life requirement is reduced will enable scientific and engineering improvements in blade technology to be incorporated on an incremental basis over the lifetime of the turbine. Such reduced service life requirement, in conjunction with a "phased blade replacement" requirement or a similar process that allows for incremental upgrades, may also potentially reduce the technology and O&M risks of poor blade performance or blade failure toward the final years of a turbine's 20 year service life.

From a science and technology perspective, as well as a public policy perspective, dissecting and isolating the legal elements from the scientific elements of *MT Højgaard* will enable the U.S. offshore wind industry to benefit from future scientific breakthroughs in engineering design by applying the lessons learned from this case. Public buy-in and positive perception about offshore wind turbines will be a critical element in the successful launch of the U.S. offshore wind industry. Neither a developer's gamble that hurricanes will not strike its wind farm, nor a designer/contractor's gamble that it will be able to create blades that will survive hurricanes and perform well for at least 20 years, are the types of bets that a cautious, risk-adverse American public wants to see its offshore wind industry take at the outset. Rather, having both developer and designer/contractors each take measures that exhibit good industry practice based on publicly available data with respect to hurricanes are things the American public likely would prefer to see. As a matter of policy, identifying steps that can be taken to both advance the U.S. offshore wind industry and improve public perception of such industry should be encouraged and taken. Industry promotion of the use of contractual provisions that allow for incremental blade replacements and updates will satisfy such dual purposes. Throwing off the shackles of a minimum 20 year service life "fitness for purpose" turbine blade provision will enable the U.S. offshore wind industry to extricate itself from the confines of a restrictive and potentially unattainable requirement. The U.S. offshore wind industry's standards are still evolving and are currently more fluid and less rigid than in other countries, particularly with respect to how to mitigate the risks associated with hurricanes. Contractual provisions that incorporate a staggered approach to blade replacement will more fairly allocate the assumption of risks associated with blade performance, provide for a certain degree of flexibility with respect to engineering and design innovations, and allow for fluidity in engineering and scientific designs while still maintaining adherence to the contracting parties' obligations. Innovations in contract provisions, as well as in scientific designs and standards, will therefore enable the U.S. offshore wind industry to advance itself toward a brighter future.

REFERENCES

4C Offshore. "Global Offshore Wind Farms Database – Robin Rigg." Available at http://www.4coffshore.com/offshorewind/index.html? lat= 54.756&lon=-3.710&wfid=UK20.

4C Offshore. "Robin Rigg – Developers/Owners/Operators." Available at http://www.4coffshore.com/windfarms/robin-rigg-united-kingdom-uk20.html.

Albiker, Johannes. "The Behavior of Monopile Foundations in Sandy Seabed Under Cyclic Horizontal Loading." Graduate School Multiscale Methods for Interface Coupling (MUSIC), Leibniz Universität Hannover. Available at http://www.music.uni-hannover.de/364.html.

Alewine, Kevin. "Wind Turbine Generator Maintenance: What to Expect and Why." *Power*, Vol 158, No. 2, at 34 (Dec. 2014).

American Wind Energy Association (AWEA). "AWEA Large Turbine Compliance Guidelines: AWEA Offshore Compliance Recommended Practices (2012): Recommended Practices for Design, Deployment, and Operation of Offshore Wind Turbines in the United States." (Sept. 16, 2012).

American Wind Energy Association (AWEA). "AWEA U.S. Wind Industry Annual Market Report Year Ending 2013." (2013). Available at http://www.awea.org/AnnualMarketReport.aspx?ItemNumber=6317.

Anatec. "Burbo Bank Offshore Wind Farm – Navigational Risk Assessment." (Aug. 2, 2002).

Atlantic County Utilities Authority (ACUA). "Atlantic City Windfarm Withstands Force of Hurricane Sandy." (Nov. 2, 2012). Available at http://www.acua.com/newsItem.aspx?id=4507&terms=sandy.

Atlantic County Utilities Authority (ACUA). "Jersey-Atlantic Wind Farm Fact Sheet." Available at http://www.acua.com/content.aspx?id=492.

Avery, Susan. *Climate Science* (C-SPAN broadcast Aug. 11, 2014).

Barkoff, Kendra. "Salazar, Chu Announce Major Offshore Wind Initiatives." Press Release. Available at http://www.doi.gov/news/pressreleases/Salazar-Chu-Announce-Major-Offshore-Wind-Initiatives.cfm.

Bell, Jim. "Building Firm Foundations for Offshore Wind." *Sea Technology*, Vol. 55, Issue 9 (Sept. 1, 2014).

Bertnes, Hakon; Leirgulen, Svein Inge. "DNV – Joint Industry Project to Develop an Improved Design Basis for the Use of Shear Keys." (March 16, 2011). Available at http://www.dnv.com/press_area/press_releases/2011/jointindustryprojecttodevelopanimproveddesignbasisfortheuseofshearkeys.asp.

"Black Swan." Available at http://www.investopedia.com/terms/b/blackswan.asp.

"Black Swan: About – What are Black Swan Events?" Available at http://blackswanevents.org/?page_id=26.

"Black Swan Theory." Available at http://en.wikipedia.org/wiki/ Black_swan_theory.

Brownen, Janine. "Wind Turbines Survive Hurricanes: What a Blow to Doubters." (Jan. 20, 2014). Available at http://yes2renewables.org/2014/01/20/wind-turbines-survive-hurricanes-what-a-blow-to-doubters/.

Bureau of Ocean Energy Management (BOEM). "Fact Sheet: Renewable Energy on the Outer Continental Shelf." (Nov. 2014). Available at http://www.boem.gov/BOEM-Overview-Renewable-Energy/.

Bureau of Ocean Energy Management (BOEM). "Offshore Wind Energy – Commercial Offshore Wind Energy Generation." Available at http://www.boem.gov/Renewable-Energy-Program/Renewable-Energy-Guide/Offshore-Wind-Energy.aspx.

Bureau of Ocean Energy Management (BOEM). "Offshore Wind Energy – Offshore Wind Energy Technology." Available at http://www.boem.gov/Renewable-Energy-Program/Renewable-Energy-Guide/Offshore-Wind-Energy.aspx.

Bureau of Ocean Energy Management, Regulation and Enforcement (BOEMRE). Available at http://www.boemre.gov/.

Cesari, Prof. Ing. FG; Mostacci, D. "Offshore Wind Energy Development in the Adriatic Sea: the P.O.W.E.R.E.D. Project as Planning Policy." *Presented at Università Politecnica delle Marche, Ancona* (May 2013). Available at http://www.powered-ipa.it/wp-content/uploads/2013/06/6_Cesari_DIENCA.pdf.

Dedić, Nedžad. "Analysis of Grouted Connection in Monopile Wind Turbine Foundations Subjected to Horizontal Load Transfer." *Institute of Mechanical Engineering, Aalborg University*. (Summer 2009).

Deign, Jason. "Monopile Failures Put Grout in Doubt." *Wind Energy Update*. (March 18, 2011). Available at http://social.windenergyupdate.com/offshore/monopile-failures-put-grout-doubt.

Diamond, Kimberly E. "Extreme Weather Impacts on Offshore Wind Turbines: Lessons Learned." *Natural Resources & Environment*. (Fall 2012).

Diamond, Kimberly E.; Scarlett, Maria. "Mitigating Shipping and Navigational Risks Around Offshore Wind Turbines: Lessons Learned from the UK." Poster Presentation at the American Wind Energy Association's Offshore WINDPOWER Conference, Baltimore, Maryland (Oct. 2011).

DNV. "DNV GL – About NNV GL." Available at http://dnvgl.com/about-dnvgl/default.aspx.

European Wind Energy Association. "Deep Water: The Next Step for Offshore Wind Energy." (July 2013). Available at http://www.ewea.org/fileadmin/files/library/publications/reports/Deep_Water.pdf.

"European Offshore 2020: Contracts – Be Prepared for New Technology Defects." *Windpower* (Nov. 8, 2013). 2013 WLNR 28119904.

"Findings from I. Lotsberg and Co-Researchers Advance Knowledge in Marine Biology." *Ecology, Environment & Conservation.* (July 19, 2013). 2013 WLNR 16849258.

"Foundations: Jackets – Jacket Required for Harsh Marine Locations." *Windpower* (Sept. 9, 2013). 2013 WLNR 22434624.

Garrad Hassan. "Garrad Hassan is now DNV GL." Available at http://www.gl-garradhassan.com/en/garrad-hassan.php.

GL Garrad Hassan. "Market Report: European Offshore Wind Industry." (June 2010).

Golightly, Dr. CR. "Monopile Grouted Connection Failures." (April 11, 2011). Available at http://www.academia.edu/6616972/Offshore_Wind_Monopile_Grouted_Connections_December_2011.

Graf, Stephen. "VP 100: New Facility in Boston to Test Large-Scale Wind Blades." (July 23, 2010). Available at http://energy.gov/articles/vp-100-new-facility-boston-test-large-scale-wind-blades.

Holland, Greg. "Extreme Impacts and Climate Variability and Change." *Climate Science* (C-SPAN broadcast Aug. 11, 2014).

International Organization for Standardization. "ISO 19902:2007 – Petroleum and Natural Gas Industries – Fixed Steel Offshore Structures." Available at http://www.iso.org/iso/catalogue_detail.htm?csnumber=27507.

Keystone Engineering, Inc. "Researchers Submit Patent Application, 'Grouted Cylindrical Connection Utilizing Bearing Surfaces for Offshore Monopile Foundations,' for Approval." *Journal of Engineering* (July 10, 2013). 2013 WLNR 16283036.

Klose, Marcus; Mittelstaedt, Marc; Mulve, Amol. "Grouted Connections – Offshore Standards Driven by the Wind Industry." *Proceedings of the Twenty-second (2012) International Offshore and Polar Engineering Conference, Rhodes, Greece* (June 17-22, 2012).

LikeForex.com. "Currency Converter Calculator". Available at http://www.likeforex.com/currency-converter/.

Massachusetts Clean Energy Center (MassCEC). "Wind Technology Testing Center." Available at http://www.masscec.com/wttc.

Minerals Management Service (MMS), Renewable Energy and Alternate Use Program, U.S. Department of the Interior. "Technology White Paper on Wind Energy Potential on the U.S. Outer Continental Shelf." (May 2006). Available at http://www.boem.gov/uploadedFiles/ BOEM/Renewable_Energy_Program/Renewable_Energy_Guide/Technology White Paper on Wind Energy Potential on the OCS.pdf .

Minns, Dr. Ned. "Challenges in the Construction of Offshore Wind Structures." (2013). Available at http://www.itpower.co.in/wp-content/uploads/2013/Ned-Minns-ITP-UK.pdf.

MT Højgaard a/s v. E.ON Climate and Renewables UK Robin Rigg East Limited, et al., Case No: HT-12-148, In the High Court of Justice, Queen's Bench Division, Technology and Construction Court, [2014] EWHC 1088 (TCC) (April 15, 2014).

Musial, Walter; Ram, Bonnie. "Large-Scale Offshore Wind Power in the United States: Assessment of Opportunities and Barriers." National Renewable Energy Laboratory (NREL). (Sept. 2010).

National Hurricane Center. "Past Track Seasonal Maps." Available at www.nhc.noaa.gov/data/.

National Hurricane Center. "Saffir-Simpson Hurricane Wind Scale." Available at http/www.nhc.noaa.gov/aboutsshws.php.

National Hurricane Center. "Tropical Cyclone Climatology." Available at http://www.nhc.noaa.gov/climo/.

Oakes, Bob; Kathleen, McNerney. "Charlestown Turbine Testing Attracts Energy Companies to Mass." (May 20, 2013). Available at http://www.wbur.org/2013/05/20/charlestown-wind-turbine-testing.

O'Brien, Leanne. "Turbine Testing Facility Opens in Charlestown." (May 19, 2011). Available at http://www.masscec.com/news/turbine-testing-facility-opens-charlestown.

"Offshore Wind Standards Due for Update." *Project Finance International* (May 6, 2010). 2010 WLNR 9313742.

Patel, Sonal. "London Array Offshore Wind Farm, Outer Thames Estuary, UK." *Power*, Vol 158, No. 2, at 30 (Dec. 2014).

Patrick, Governor Deval. Press Release. "Secretary Chu, Governor Patrick Announce $25 Million for Massachusetts Wind Technology Testing Center." (May 12, 2009). Available at http://www.mass.gov/governor/pressoffice/pressreleases/2009/25-million-for-massachusetts-wind-technology.html.

Price, Toby. "Grouting Still a Major Issue for Offshore Wind." *Renewable Energy Magazine*. (Jan. 4, 2012). Available at http://renewableenergymagazine.com/article/grouting-still-a-major-issue-for-offshore.

Quarton, David. "An International Design Standard for Offshore Wind Turbines: IEC 61400-3". (2005). Available at http://wind.nrel.gov/public/SeaCon/Proceedings/Copenhagen.Offshore.Wind.2005/documents/papers/Design_basis/D.Quarton_pp.pdf.

Rainy Sky SA v. Kookmin Bank, Supreme Court, [2011] 1 W.L.R. 2900 (July 27, 2011).

Segeren, Maxim. "Wind Power Innovations." *European Wind Energy Association, Annual Conference.* (Feb. 5, 2013). Available at http://proceedings.ewea.org/annual2013/programme-final/info2.php?id2=942&id=168.

"Sheringham Bearings Fitted." *Platts Power in Europe* (Oct. 31, 2011). 2011 WLNR 23572127.

"Solway Firth." Available at http://en.wikipedia.org/wiki/Solway_Firth.

"Subsea O&M: Foundations Built on Corrosion Protection." *Windpower* (Sept. 30, 2014). 2014 WLNR 27200297.

Swaney, Chriss. "Press Release: Carnegie Mellon Team Finds Hurricanes Pose Potential Risks to Offshore Wind Turbines." Carnegie Mellon University. (Feb. 14, 2012). Available at http://www.cmu.edu/news/stories/archives/2012/february/feb14_windturbinesatrisk.html.

The Crown Estate. "A Guide to an Offshore Wind Farm." Published on behalf of The Crown Estate.

UK Energy Research Centre (UKERC). "Great Expectations: The Cost of Offshore Wind in UK Waters – Understanding the Past and Projecting the Future." (Sept. 2010).

U.S. Department of Energy (DOE), Office of Energy Efficiency & Renewable Energy, Wind & Water Power Program. "A National Offshore Wind Strategy: Creating an Offshore Wind Energy Industry in the United States." (Feb. 7, 2011). Available at http://www1.eere.energy.gov/wind/pdfs/national_offshore_wind_strategy.pdf.

U.S. Department of the Interior (DOI) Bureau of Land Management, DOI-BLM-CA-060-0007-0057-EA. "Right-of-Way (R/W) Grant CACA-11688-A, Mesa Wind Power Corporation." Palm Springs-South Coast Field Office. (Feb. 13, 2008).

World Meteorological Organization. "Press Release No. 943: WMO Annual Statement Confirms 2011 as 11th Warmest on Record; Climate Change Accelerated in 2001 – 2010, According to Preliminary Assessment." (March 23, 2012). Available at https://www.wmo.int/pages/ mediacentre/press_releases/pr_943_en.html.

X-Rates. "Currency Calculator." Available at http://www.x-rates.com/calculator/?from=EUR&to=USD&amount=26,250,000.

In: Offshore Wind
Editor: Chester Mendoza

ISBN: 978-1-63482-364-7

Chapter 3

OFFSHORE WIND: SYSTEM-LEVEL VIABILITY?

M. J. Platts*
Institute for Manufacturing, University of Cambridge, UK

ABSTRACT

Offshore wind turbine rotors and generators create value by 'farming' the wind, converting its energy into electricity. Costly infrastructure – the very high towers, deep seabed foundations, buried high voltage insulated cables, on-platform transformers and complete subsea transmission grid – and expensive at-sea maintenance are required because this electricity is not where it is needed. The cost of this infrastructure cripples the investment payback ratio. High investment across the world in specialised manufacturing plant and tens of thousands of experienced workers enable the rhythm and flow that mass produces some 20,000 wind turbines a year of the now-standardised designs of wind turbines erected on land close to where the energy is needed, driving their cost effectiveness. Offshore represents a 2.5% extension of the market, in disjointed projects, requiring the expensive development of much larger wind turbines and requiring yet more investment in much larger factories but offering no chance of continuous rhythm-and-flow production and also requiring the development of much new and expensive offshore infrastructure-building technology to make it possible

* Corresponding author: mjp@eng.cam.ac.uk.

at all. This kind of 'bigger' is not 'better'. Technical heroics is no substitute for the lean and agile thinking within the industry itself that is steadily developing lighter, larger, more dynamically responsive rotors and lighter, taller towers, enabling today's standardised wind turbines to cost effectively 'farm' the lighter winds close to where the electricity is needed, with minimal infrastructure costs.

INTRODUCTION

There are three fundamental laws that have to be understood concerning energy infrastructure development. They cover understanding, technical competence and relationship competence.

Firstly, an energy resource may have a potential value, but it only has an actual value when that energy has been delivered to the user's point of need. The cost of the infrastructure needed to get it there may negate the whole potential value of the resource.

Secondly, research and development may demonstrate a technology that has potential value. It does not have actual value until the complex interdependent competencies involved in manufacturing it reliably, in quantity, to a timescale and at a viable cost have been established.

Thirdly, wilfulness begets neither understanding nor competence. In fact it demonstrates an absence of both. The skill of civilisation itself is the skill of being able to listen to each other, understand each other's needs and skills and work together in a coordinated, responsible and trustful way to understand what is possible and to achieve a collective outcome that has both value and meaning and is treasured by all. This is a shared exploration and growth process and has to be managed as such. You cannot solve this kind of problem by presuming the answer before you begin.

LOCATION V VALUE

Wind energy is a distributed natural resource that has to be farmed. It is farmed by the aerodynamics of rotating wind turbine blades that convert the flowing energy in the wind to rotating mechanical energy in a shaft. This is then converted to electricity by a combination of a gearbox, an electrical generator, power electronics and a control system. All these add value, but the value is only realised when the electricity is delivered to a customer at their point of use. An electricity infrastructure is needed to transport the electricity from the point of generation to the point of use, and this adds cost. It may or may not add value.

It is helpful to think separately about 'close' location and 'far' location. One part of this infrastructure, which is commonly thought of as part of the wind turbine itself, is the tower holding the turbine up in the air. The amount of energy in the wind increases with the wind speed and the speed of the wind increases with height from the ground, so a taller tower will cost more, but it will also add value by moving the turbine into a higher wind speed location. This is a 'close' location issue, and since the design and manufacture of the towers are entirely within the control of the wind turbine manufacturers themselves, they can readily achieve optimum value at many different locations.

'Far' location decisions are a much more difficult problem, involving potential wind farm investors and operators trying to assess the potential value of putting wind turbines in higher wind speed locations, compared to the cost of building the infrastructure needed to bring the electricity to the customer's point of use. This will involve planning, costing and in due course construction by many different bodies, many not directly involved with wind energy at all, and will have impact on many other bodies not involved in the basic transaction and so it is very difficult to assess the true costs and allocate them appropriately. This is the nature of all infrastructure investment and is a subject worthy of coordinated research. 'Infrastructure' is a truly 'socially owned' product involving many aspects that are not at all simple interactions representable by buy-sell transactions in a market.

Far Location Problems in Germany

By 2015 [1] Germany had over 35GW of wind energy capacity installed onshore and 922MW offshore, but Germany has problems [2]. Over two thirds of its wind energy capacity is in North West Germany, some 1000km from the main load centres in South East Germany. Grid development has not been keeping pace with the rate of installation of this new generation capacity nor is upgrading the grid a simple task. Much of the grid was built decades ago as part of a very simply structured 'top down' electricity supply system, beginning with large power stations generating at high voltage and radiating outwards through middle voltage one way distribution systems to low voltage consumer delivery networks. Wind farms represent low or medium voltage dispersed generation sources wanting to send electricity the other way. Not only does this overload the network, it creates flow requirements and monitoring, switching and control requirements that never existed before, all over the grid. Simulating what might be required and exploring different design options is a major task in itself, let alone organising and doing it all and identifying who pays for what and why. Germany has to route some of its NW-SE power flows out of Germany, through the grids of neighbouring countries and back into Germany to maintain its supply. So solving Germany's grid development problem is already involving European-level cooperation.

Subsea Foundations Sink Energy Payback Ratios

Looking at pictures of offshore wind turbines they look very like wind turbines onshore. However, there is much that is expensive under the sea. Taking as an example the 3.6MW Siemens turbines on the UK Walney wind farm, in 30m water depth, each above-water wind turbine and tower weighs 480 tonnes but it requires 900 tonnes of subsea steel in the cylinder that goes into the sea bed and in the connector piece that links it to the tower. In the initial stages of any development it is always possible to claim that the costs are unrepresentative and they will improve. What is a far more telling figure, because it does not change in quantity production is to look at the energy investment represented by the material embodied in the wind turbine, tower and subsea supporting structure, to establish whether this is likely to be a good investment. When what is being built is supposedly a sustainable energy source, the payback ratio in energy terms is an important figure to be aware of.

Wind turbines usually have a 20 year life and, at a good location onshore a wind turbine in the 2MW range will usually demonstrate an energy payback ratio over their lifetime of around 40:1 (i.e., they have 'paid back' the energy invested in making, installing and maintaining them through their operational lives, in the first six months of operation and the remaining 19.5 years of output is genuinely environmentally clean energy 'profit'), making them an attractive source of sustainable energy. It is important to note however that 20 years is a very short lifetime for something intended to be 'infrastructure' and the need for their replacement after 20 years is often overlooked in discussions concerning energy policy. This ratio would of course improve if further development could give them a longer operational life but moving to locations giving more extreme loadings on the wind turbines weighs against this. In her examination of the energy payback ratios of offshore wind turbines, Arrou-

Vignod [3] estimated an energy payback ratio for the 3.6MW Siemens wind turbine offshore in the above-described installation of 15:1, a significant reduction in its 'sustainable energy' productivity because of the extra energy embodied in all the steel in its foundations. Because of the heavy fatigue damage suffered by the foundations as well as the wind turbine itself in these sorts of turbulent locations, the whole installation has a 20 year life and needs to be replaced after this time.

But this is only the wind turbine part of the story. What happens in taking the electricity grid offshore?

Electricity Transmission Costs All at Sea

By 2013 [4] Siemens had completed four offshore wind farms – of 576, 690, 800 and 864MW, with bespoke transformers and cabling for each, running at 250, 300, 320 and 320KV, because these were all separate projects running to uncoordinated schedules with no continuity of supply chain work possible between them. And these were all close inshore AC transmission projects. They did not involve the much more complicated conversion to high voltage DC transmission required for wind farms being planned much further out to sea. But they still involved substantial sub-sea work on major seabed foundations for each turbine, similar to that outlined above, and all transmission cabling deeply trenched into the seabed. This is expensive electricity - €128-142/MWh compared to €60-80/MWh from onshore wind turbines - and Germany has paused in its offshore wind development work. Its forward cost projections for these close inshore wind farms is €91/MWh by 2023 and €100/MWh for far offshore [5], with the wind turbines themselves representing only 31.8% of the costs. A comparable US study suggests costs 12% higher than this.

Development vs. Manufacturing

As of December 2014, the global industry's forward pipeline of wind farm projects in planning amounted to 460GW, of which around 300GW will probably be built. Closer to the present, the forward order books of the ten largest wind turbine manufacturing companies globally – indicative of a two year forward view – totalled 58.5GW capacity, around 30GW per year, which will mostly be onshore wind farms using their well established wind turbine designs in the 2-3MW range, of 90-110m diameter, the companies collectively manufacturing and installing around 12,000 wind turbines a year [6]. Vestas and Siemens, the two largest wind energy companies globally, each have around 12GW on their forward order books, representing around 6GW each a year – over 2000 wind turbines of this size a year, meaning a manufacturing rate of some ten wind turbines a day each. Wind energy represented $76bn in infrastructural investment globally in 2013.

Henrik Stiesdal, retiring from being CTO of Siemens Wind Power after 35 years in the industry [7], considers that the core of the industry is now well developed and future progress is mostly going to come from steady evolution of the technology, not from any kind of sudden transformational change, primarily because the depth of understanding of the total system needed to get every bit right is now so high that no outsiders are likely to understand enough to be able to contribute well to the overall story. An assessment of outside projects hoping to offer improvements [8] shows levels of funding totalling around 0.1% of the industry's own internal annual spend and thus far less than its own internal annual research, development and demonstration budget.

Integrated Supply Chains

The major companies in the wind energy industry head supply chains that have been together for over thirty years, developing together the technologies, the skilled workforces and the very substantial factories needed to make the blades, the gearboxes, the generators, the high voltage power systems, the control systems, the towers, and the large mobile cranes to assemble these turbines on site, all working in synchronised rhythm at a steady rate of production. Over 200,000 people are involved globally and they are rightly proud of what they achieve every day. It takes many years to develop this coordinated ability and it is this ability that produces high reliability, high value products cost effectively.

Every manufacturer has developed what are known as 'platform technologies' – core technologies that support a family of turbines within a specific size range, all manufacturable in the same factories on the same machines, with the variations allowing the manufacturers to give optimum value for the different detailed requirements of different customers, each with their specific locations and wind conditions. The flow of wind turbines from these factories and the experiences of them operating in all these different locations also provide a constant learning process that enables the manufacturers to continuously improve details in all the components and in the overall manufacturing process, so that the designs constantly evolve and improve in their reliability and cost effectiveness.

Developing a new 'platform technology' for a physically much larger wind turbine is a major undertaking, not only taking large teams many years to evolve the design and make one-off components for testing and then make a complete wind turbine for testing. Because of its larger size this new 'platform technology' cannot be manufactured in the existing factories by the existing skilled workforce (who are all fully occupied anyway earning the company its living). It requires the coordinated investment in a new set of factories and in the recruitment and training of new workforces, to initiate manufacturing at this new scale. Creating this new manufacturing base is a major forward investment far greater than the cost of development of the technology itself – ten to a hundred times as much - and requires clear forward sight of an order book large enough and firm enough to warrant the investment.

When these larger scale technologies are for quite fundamental reasons less cost effective than the existing technology (there is a fundamental square/cube law that mitigates against size in wind turbines) and the projects that involve these large turbines are themselves in difficult locations and are hard to develop, so that only a staccato pattern of projects appears possible, this is not a promising scenario. At its global blade technology development centre on the Isle of Wight in the UK, Vestas has made and tested an 80m blade and made three blades for its prototype V164 turbine. Its holding position is that this small but very skilled team could make more 80m blades if needed, at no further investment [9]. This would be one blade every few days, not a few blades every day. This is 'making', not 'manufacturing'.

Light and Flexible – Close Location Success

Where the industry's own R&D is going is in continued improvement in the cost effectiveness of wind turbines for the light wind conditions onshore in areas close to the major load centres and where most of the requisite electrical power infrastructure already exists. The Nordex N131/3000 turbine completes Nordex's 'Delta platform' range, with a 3MW turbine designed for low wind IEC IIIA locations, with a 131m diameter rotor specifically designed to be quiet, with an in-house developed 64.4m very slender blade making use of 'flat-back' airfoils further towards the root to maximise energy capture whilst minimising blade loading [10].

Enercon is also developing a new 4MW 'platform technology', but again, not for use offshore. It is similarly focused on exploiting low wind speed areas more cost effectively that is currently possible and, crucially, its design life is 30 years, not 20 years. The development of the EP4 platform has already taken a permanently engaged team of 20 experts two years and other experts from the company's 500-strong R&D department have also steadily contributed specialist inputs. The foci have been cost-effective large-series manufacture, optimised transport logistics and uncomplicated site assembly and installation. The direct drive generator is designed to be manufactured in two 180-degree segments, enabling train transportation and optimised road transportation logistics. New stator coil designs enhance heat dissipation, enabling air cooling, reducing cooling system energy consumption by 35% and reducing noise. The new, quieter, long and slender E-126-EP4 blade design, also making extensive use of 'flat-back' airfoils to maximise energy capture whilst reducing loads by 30% thus minimising fatigue load input, is similarly designed for cost-effective transportation, easy assembly on site and long life. Enercon have developed in-house slim concrete-steel towers offering hub heights up to 144m for the EP4 family of turbines, and the power electronics in the tower base include a split transformer design to optimise upkeep. Whilst their currently very successful smaller wind turbines will continue to be their mainstay for some years, they expect this family of larger turbines to become a significant part of their production in the future. The prototype will begin testing in late 2015 and series production will begin in 2016 [11],

CARE AND GROWTH VS. MANAGEMENT BY OBJECTIVES

The *name* of the Institution of Civil Engineers – the oldest professional engineering body in the world, founded in 1818 and given its Royal Charter in 1821 – embodies a key insight. They called themselves *civil* engineers because they were *not* military engineers. What they realised was that when you have achieved the ability to coordinate the work of thousands of men spread over hundreds of miles and working over many years, to build a canal system or a railway, in a way that is transparent and fair to all, so all get fair pay for fair work and all can see what they have achieved and can be proud of what they have done… the *product* is *infrastructure*, but the *process* is *the process of civilisation itself*. It is *not* a process of confrontation. It is a process of collaboration, of shared struggle, shared insight and shared success.

Infrastructure is by its very nature long term and completely pervasive and it requires full civic understanding, collaboration and ownership, and full civic commitment and investment for its achievement. Saying that you want it, without doing enough homework to understand what is involved in creating a real, valuable and long lasting solution to an as yet not fully understood need, is childish in the extreme, a complete externalisation of responsibility which manifests as the issuing of un-thought-out orders and then allocating blame by pointing fingers when it turns out to be more complicated than that and not achievable as desired. In cases like this the responsibility actually lies at home. Emotionally driven politics is as bad a driver of any 'development project' as emotionally driven engineering. Real leadership involves caring deeply enough to develop a good operational understanding, not only of the totality of the need but of the social and technical processes that need to be developed and sustained for the development of a viable future. It is these processes that really need the research and development. If they are done right, the technology will look after itself.

Politically Created Markets

The function of a market is as a communication system in which price communicates value, and terms and conditions communicate relationships. Unfortunately no energy market is transparent. All have political subsidies and restrictions placed on them, which distort them. When a market is being politically created it requires a stable political price to be set. There appears to be a steady retreat from exploratory investment in UK offshore wind farms because of a failure of this process. Centrica withdrew from the £2b offshore Norfolk Race Bank wind farm project in 2013 citing failure to secure adequate consumer-funded subsidies for it and, with their co-developers, also closed the Celtic Array offshore wind farm project as 'uneconomic' taking a £40m write-down on it. In 2014 they also sold their stake in the Barrow project to Dong Energy, so Centrica has no forward offshore wind farms in its pipeline [12].

A market can also be destroyed politically. Dale Vince, owner of Ecotricity, a major UK wind farm investor, has withdrawn from onshore wind development explaining that the planning process has become unworkable, with terms of planning, environmental assessments and financial support changing every few months. The Government Minister concerned (Mr Pickles) has intervened in the planning of 53 wind farm projects, calling in the applications so that he can make the final decision. In 25 decisions so far, 22 have been rejected, in 7 cases overruling the recommendation of an independent inspector that the wind farm should be built [13].

Corrupting the communal communication process in this way - by reaching for control in the face of difficulty - is a generic problem identified by Lord Acton in 1887. In a letter to Bishop Creighton, 5 April, he wrote: “Power tends to corrupt, and absolute power corrupts absolutely. Great men are almost always bad men… There is no worse heresy than that the office sanctifies the holder of it.”

Ceri Galloway, a Green Party candidate for local government, in a letter in the Cambridge News concerning local planning issues, 31 December 2014 [14], writes very precisely about the source of this unhelpful behaviour being a lack of a developed competence of how to catalyse fruitful collaborative interchanges and thus a lack of confidence that such a process is possible. It is worth quoting the letter in full:

“I note that since the Labour party became the majority group in Cambridge City Council, the practice of bringing planning decisions to the local area committees no longer takes place. This is a backward step, reducing local democracy and centralising planning processes. Taking this cabinet style approach does not empower local residents.

The value of bringing planning proposals to an area committee was that it offered an opportunity for residents to discuss documentation with planners and councillors and raise their concerns. It also helped demystify the planning process. Crucially, the meetings took place in the evening when local residents were much more likely to be able to attend. Now residents are required to attend daytime meetings at the Guildhall if they wish to speak on planning applications.

Cambridge is changing at such a rate that it is hard for residents to keep up with all the planning applications affecting their local area. By taking it back in house Labour has weakened the way in which local people were able to keep in touch with what is happening.

It is important for local people to have a say on planning issues that affect them directly. In return they can offer useful and detailed local knowledge. This two way process is beneficial to both planners, developers and the community leading to higher standards in applications.

At the time I was vice-chair of the Trumpington Residents Association, for instance, the local Residents Association became a partner in the planning process for a project that has brought long term benefits to the community, and the input by the association meant that the planning application was of much higher standard than the initial application. Surely the involvement of community champions with local knowledge is preferable.

Observations made at a recent community meeting about planning indicated that the Labour councillor chairing the meeting lacked knowledge of how to engage the public in council debates and how to empower them to participate. He appeared uncomfortable and when asked about planning further meetings of this type, he seemed to lack understanding about the benefits of this type of meeting.

As Greens, we encourage local democracy and active decision making. We want councillors to draw on local knowledge, not only because it brings a community benefit, but because the loss of the open forum is a step backwards for local democracy."

In comparing Spain's development of a globally competent wind energy industry and a large quantity of wind energy capacity on its grid, with the UK's failure to do either over the same thirty year period, Chang Yong Ahn [15] highlights this civic procedural competence and confidence as the source of Spain's success and the lack of it as the source of the UK's failure. These are not small issues.

CONCLUSION

'Infrastructure' is in some ways not a 'market' at all. It is the single greatest gift that any generation bequeaths to its grandchildren. Its ubiquitous, reliable presence is testimony to the magnitude of the gift and being able to give it, generation after generation, is civilisation's greatest test. Achieving it requires the cohesive investing of the skills to create it, the will to finance it and the sustaining of transparent rules-of-the-game that engender shared insight, understanding and commitment and are fair to all. Building prototype wind farms and electricity systems that will only last 20 years, in the North Sea, far from where the electricity is needed, might be heroic engineering, but is it infrastructure?

REFERENCES

[1] Windicator. (2015). 'Global Index', *Windpower Monthly*, January 2015, p. 46.

[2] Robb, D. (2014). 'Helping Germany's transmission grid keep up with wind capacity growth', *WindStats Report*, Q4 2014, 4-6.

[3] Arrou-Vignod, A. S. (2012). *Energy Balance Estimation of Offshore Guyed Tower Wind Turbines*, MPhil Thesis, IfM, University of Cambridge.

[4] Knight, S. (2013). 'Cabling standards hold key to cutting costs', *Germany Offshore – Windpower Special Report*, April 2013, 7-8.

[5] Milborrow, D. (2014). 'Offshore costs: Cost of wind continues to fall', *Windpower Monthly*, November 2014, p. 45.

[6] Campbell, S. (2014). 'Concentration on consolidation'. *Windpower Monthly*, December 2014, p. 22.

[7] Quilter, J. (2014). 'Steisdal on new records, old turbines and growth', *Windpower Monthly*, December 2014, p. 48.

[8] Davidson, R. (2014). 'High-profile failures cast doubt on VC funding', *Windpower Monthly*, December 2014, p. 11.

[9] Quiltier, J. (2014). 'MHI-Vestas £200 million UK plan remains short on detail', *Windpower Monthly*, December 2014, p. 19.

[10] Vries, E. D. & Buist, J. (2015). 'The best wind turbine products of 2014', *Windpower Monthly*, January 2015, 20-28.

[11] Vries, E. D. (2015), 'Enercon unveils 4MW platform with 30-year lifespan'. *Windpower Monthly*, January 2015, 37-38.

[12] Webb, T. (2014). 'Wind stops blowing for Centrica, *The Times*, 20 December 2014, p. 50.

[13] Webster, B. (2014). 'Wind farms owner gives up over political interference', *The Times*, 25 November 2014, p. 6.

[14] Galloway, C. (2014). 'Put plans back on the agenda', *Cambridge News*, 31 December 2014, p. 20.

[15] Ahn, C. Y. (2009). *An Exploration of the Role and Nature of Entrepreneurship in Industry Creation: A comparative Study of the Emergence of the Wind Industry Sector in Britain and Spain*, PhD Thesis, IfM, University of Cambridge.

In: Offshore Wind
Editor: Chester Mendoza

ISBN: 978-1-63482-364-7

Chapter 4

EFFECTIVE TECHNO-ECONOMIC EVALUATION METHOD FOR OFFSHORE WIND NOVEL PROPOSALS

***A. Madariaga*[1,*], *J. L. Martín*[2], *I. Zamora*[2], *S. Ceballos*[3] and *O. Anaya-Lara*[4]**

[1]Offshore Renewable Energy Catapult, Glasgow, UK
[2]University of the Basque Country, Bilbao, Spain
[3]Tecnalia Research & Innovation, Bilbao, Spain
[4]University of Strathclyde, Glasgow, UK

Abstract

This chapter presents a new method to assess techno-economically innovative solutions for offshore wind projects. The main aim of the proposed technique is to provide academic and industrial technologists with a new tool to conduct accurate cost benefit analyses in an agile way. The method proposes a simplified life-cycle economic analysis, taking into account the time value of money and the most relevant economic variables. Its structure also enables to study the effect of long term variations in wind speed, electricity price and other variables in conjunction in a simple and effective way. The assessment of the energy yield of the offshore wind power plant (OWPP) considers the electric power losses in all the systems that make up the windfarm, as well as the wake effect and the unavailability of the different systems. The formulation of the method

*E-mail address: ander.madariaga@ore.catapult.org.uk

enables it to be shaped either in a spreadsheet or in algorithmic form, avoiding power flow simulation studies up to now used by utilities and consultants to assess the electric power losses in grid integration studies. By doing so, the proposed method decreases the time needed to conduct the studies and increases staff capable of performing the task.

Keywords: offshore wind, research economics, cost-benefit analyses, grid integration studies

UNESCO Fields of Science and Technology: 3306, 3307, 3322

1. Introduction

With an installed capacity in Europe around 8 GW by the end of 2014, the offshore wind industry offers tested solutions that can be considered almost industrialized under certain conditions of sea bed depth and distance from shore. However, as in any other emerging industry, optimized solutions are far from being available and offshore wind power plants (OWPPs) that will be built during the next decade will probably have to introduce new concepts to cope with the new challenges. Among these challenges, a significant cost reduction has to be achieved in the near future to meet the installed capacity forecasts. Several institutions and organisations, particularly in the United Kingdom, are promoting strategies to achieve that objective [1, 2, 3] and industry and academia are considering new solutions for projects under development.

Within the new global energy landscape, investors considering offshore wind will surely demand to keep risks at minimum, particularly bearing in mind uncertainties related to German offshore wind development, and every industrial or academic project will definitely require a strong case to move forward, ideally with a clear cost and risk decrease potential associated with it. Therefore, techno-economic assessments and the methods used in these analyses will surely play a key role. Nonetheless, to prepare a realistic cost-benefit analyses for an offshore wind project is not an easy task because of the technical complexity of the systems and the correlation between economic, environmental and technical variables. A review of different approaches proposed during the last decade can be found in [4].

With this context in mind, the chapter presents a new method to assess techno-economically innovative solutions for offshore wind projects. The main aim of the proposed novel technique is to provide academic researchers, industrial technologists and engineers with a simple tool to conduct accurate cost benefit analyses in an agile way. The method proposes a simplified life-cycle economic analysis, taking into account the time value of money and the most relevant economic variables. The assessment of the energy yield of the OWPP considers the electric power losses in all the systems that make up the windfarm, as well as the wake effect and the unavailability of the different systems. The formulation of the method can be shaped either in a spreadsheet or in algorithmic form, avoiding power flow simulation requirements in electrical layout optioneering studies conducted in the early stages of concept design. By doing so, the proposed method decreases the time needed to conduct the studies and increases staff capable of performing the task, therefore increasing the number or technical alternatives that can be evaluated during this stage. The structure of the proposed technique also enables to study the effect of long term variations in wind speed, electricity price and other variables in conjunction in a simple and effective way.

Besides the introduction, the rest of the chapter is structured into six sections. Section 2 lists all technical and economical values considered in the formulation, pointing out the symbol, units and a short conceptual description. Section 3 presents the economic evaluation structure of the method, whereas section 4 presents the formulation of the yearly average power production assessment technique. Section 5 discusses the validation of the proposal, and section 6 provides full details on how to apply the method to a case study. Finally, section 7 concludes the chapter pointing out the key contributions of the new technique and some lines of future work.

2. Symbols

Symbol	Unit	Description
C	F / km	Cable capacitance
$CF_{t,1}$	£	Loan instalments
$CF_{t,2}$	£	Operational expenditures (OPEX)
$CF_{t,3}$	£	After tax incomes

Symbol	Unit	Description
$CP_{t,r}$	$^{\circ}/_{1}$	Cumulative probability for the wind blowing at $ws \pm 0.5$ during year t
EP_t	£/MWh	Price obtained for the electricity sold
F^{cs}_{unav}	$^{\circ}/_{1}$	Collector system unavailability factor
F^{grid}_{unav}	$^{\circ}/_{1}$	Grid unavailability factor
F^{owts}_{unav}	$^{\circ}/_{1}$	Wind turbine unavailability factor
F^{owts}_{wake}	$^{\circ}/_{1}$	Wake effect factor
F^{ts}_{unav}	$^{\circ}/_{1}$	Transmission system unavailability factor
I_{reac}	A	Phase current in offshore reactors
L	H/km	Cable inductance
$L_{ts,cab}$	km	Length of the HVAC submarine cable
N_{clu}	-	Number of clusters in the wind farm
$N_{owt,clu}$	-	Number of turbines per cluster
NPV	M£	Net present value of the project
P_{Cu}	W	Copper losses in power transformers
P_{Fe}	W	Iron losses in power transformers
P^{cs}_{in}	W	Power input to the collector system
P^{cs}_{lost}	W	Power lost in the collector system
P^{cs}_{out}	W	Power output from collector system
P^{cs}_{unav}	W	Power lost due to collector system unavailability
$P^{cs,clu}_{in}$	W	Power fed into a cluster
$P^{cs,clu}_{lost}$	W	Power lost within a cluster
$P^{cs,cab}_{lost}$	W	Power lost in MVAC cables
$P^{cs,str}_{lost}$	W	Power lost in one MVAC cable stretch
$P^{grid}_{t,r}$	MW	Power fed into the grid during the year t with turbines in r operating range
P^{grid}_{t}	MW	Average power fed into the grid
P^{grid}_{unav}	W	Power lost due to grid unavailability

Symbol	Unit	Description
P^{owpp}_{rated}	W	Rated power of the OWPP
P^{owpp}_{in}	W	Power input from all the wind turbines
P^{owpp}_{lost}	W	Power lost within the OWPP
P^{owpp}_{out}	W	Power put at onshore grid disposal
P^{owt}_{r}	W	Power output of a turbine in r range
P^{owts}_{unav}	W	Power lost due to turbines unavailability
P^{owts}_{wake}	W	Power lost due to wake effect
P^{ts}_{in}	W	Power fed into the transmission system
P^{ts}_{lost}	W	Power lost in the transmission system
P^{ts}_{out}	W	Power output from transmission system
$P^{ts,cab}_{lost}$	W	Power lost in HVAC cable
$P^{ts,ofco}_{lost}$	W	Power lost in offshore shunt reactors
$P^{ts,ofpts}_{lost}$	W	Power lost in offshore transformers
$P^{ts,ofpts}_{in}$	W	Power fed into offshore transformers
$P^{ts,onco}_{lost}$	W	Power lost in onshore STATCOM
$P^{ts,onpt}_{lost}$	W	Power lost in onshore transformer
P^{ts}_{unav}	W	Power lost due to transmission system unavailability
PF_{cs}	$^{\circ}/_{1}$	Power factor in collector system
PF_{ts}	$^{\circ}/_{1}$	Power factor in transmission system
Q	-	Reactor quality factor
R_t	£	Net cash flow (after taxes) assessed for year t discounted today
R_{reac}	Ω	Shunt reactor per phase resistance
$S^{ts,ofpts}_{rated}$	VA	Rated apparent of power offshore transformers
U_{cs}	V	Collector system voltage
U_{ts}	V	Transmission system voltage
X_{line}	Ω	HVAC cable reactance
X_C	Ω	HVAC cable capacitive reactance

Symbol	Unit	Description
X_L	Ω	HVAC cable inductive reactance
X_{reac}	Ω	Reactor per phase actual reactance
X_{reac}^{hv}	Ω	Reactor per phase reactance at HVAC side
ctr	%	Corporate tax rate
d	%	Discount factor
ws	m/s	Wind speed
$\overline{ws}$	m/s	Mean wind speed
ω	rad / s	Pulsation ($2\pi f$)

3. Economic Assessment

3.1. Selection of the Evaluation Criteria

Different evaluation methods have been used in the past to compare electric topologies in the context of offshore wind. Some of them have used the cost of energy (COE) as the comparison criteria [5, 6, 7]. Other approaches have used the levelized cost of energy (LCOE) to take into account the time value of the incomes and the costs [8, 9]. The net present value (NPV) of the investment has also been used to decide which is the best layout under certain assumptions [10].

On the other hand, there have also been analyses focused on particular aspects of offshore wind projects such as capital expenditure (CAPEX) or electric power losses [11, 12]. Other studies have presented detailed proposals to optimize the life-cycle costs of certain systems adapting the techniques used by utilities in new asset optioneering [13, 14, 15]. And during the last few years there have been a considerable number of proposals to design the electric layout of OWPPs following different optimization methods [16, 17, 18].

Therefore, the first point is to choose which evaluation criteria suits best the objective of the method. Let us consider the following statements:

1. The overall perspective should not be lost by only considering capital costs or electric power losses. The offshore wind turbine (OWT) industry has recently given a clear example. Permanent magnet synchronous generators are significantly more expensive than squirrel cage induction

generators. In spite of that, they are becoming a clear trend in multi-MW new OWTs because they enable manufacturers to simplify or even eliminate the gearbox, with a significant foreseen decrease in its operational expenditure (OPEX).

2. Non-analytic stochastic optimization methods are an effective option for finding rapid, robust and good solutions in power system operation due to the dimensions and complexity of the problem [19]. But we should bear in mind that when utilities and transmission system operators (TSOs) decide to use these methods, they have at their disposal extensive databases with the key operational data of the grids under study, and they have the possibility of checking and properly tunning these optimization methods before using them. On the contrary, extensive checks with actual data are not possible yet in the context of offshore wind, and therefore, it is unclear whether these techniques are suitable for OWPP optioneering at present time.

3. With approaches based on the COE calculated on an annual basis it is not possible to consider either the lifetime value of money or the influence of variations in the key variables throughout the lifetime of the offshore windfarm.

4. The LCOE considers the time value of money and is usually aimed at reporting the cost of electric generation with different technologies in consultancy reports, technology roadmaps and project tenders. Therefore, it is a metric well known to policy makers, consultants and project managers. However, it has been decided not to use it because of the lack of a widely accepted definition. In [20] for example, the LCOE is defined as the total revenue required per MWh of energy, so that the windfarm owner secures a 10 % return on CAPEX and OPEX paid from its balance sheet over the project lifetime[1]. Similar definitions have been apparently used in other institutional reports in the UK [22, 23], although no explicit mention appears in them. On the contrary, in [24] a completely different approach is applied, updating the definition previously presented by NREL in [8].

[1]It makes sense to note that funding schemes for oncoming offshore wind project are expected to change from the private equity and venture capital schemes of the last decade to funding strategies mainly based on public equity markets [21]. Therefore, it is not clear whether an approach that considers developers' equities as the main funding source is appropriate.

Bearing all this in mind, the net present value (NPV) of the offshore wind project, obtained by means of a discounted cash flow (DCF) procedure, has been the approach preferred in the new method proposed. When applied to the comparison of alternative OWPP electric topologies, it gives information about the overall economic impact of a certain technological decision. The LCOE of the offshore wind project under evaluation could also be implemented as an output of the method provided that a precise definition is given, because its calculation shares many characteristics with that of the NPV.

3.2. The Net Present Value Assessment

In finance, the NPV or net present worth of a time series of cash flows, both incoming and outgoing, is defined as the sum of the present values of the individual cash flows of the same entity. The formulation used in this methodology is presented in (1).

$$\begin{aligned} NPV &= \sum_{t=1}^{lifetime} R_t \\ R_t &= \sum_{k=1}^{3} CF_{t,k}\,(1-d)^t\,(1-ctr) \end{aligned} \tag{1}$$

The amount of the loan instalments ($CF_{t,1}$) is calculated considering the CAPEX of the project, and assuming that the financing scheme is a loan obtained in public equity markets, with constant interest rate and fees due at the end of each year [2]. Regarding the OPEX ($CF_{t,2}$), the structure of the NPV evaluation procedure enables to consider different values throughout the lifetime of the asset. However, the determination of a realistic evolution of the offshore windfarm OPEX is outside the scope of the work presented in this chapter. The income obtained from the electricity sold in the market during each year ($CF_{t,3}$) is calculated using (2) considering the corresponding electricity price (EP_t) and average power fed into the grid (P_t^{grid}), which is assessed with the formulation detailed in section 4.

$$CF_{t,3} = 8760\, P_t^{grid}\, EP_t \tag{2}$$

[2]Considering the amount of variables in the formulation their description has not been always embedded in the text to improve the readability of the chapter. A comprehensive list has been provided in section 2.

The approach is simple in economic terms bearing in mind current financing schemes with up to ten private and public investors [25]. However, it conducts a life-cycle economic assessment taking into account most of the variables that influence these analyses, *i.e.* the lifetime of the OWPP, the CAPEX, the installments at a certain interest rate and reimbursement period, the OPEX, the foreseen energy yield, the electricity prices, the corporate tax rate and the discount factor to assess the time value of money. Moreover, the influence of long-term variations in wind speeds, electricity prices and OPEX can be considered in conjunction to assess some of the risks associated with offshore wind projects.

4. Electric Power Production Assessment

The varying natures of the power source and of the response of the OWPP in terms of electric power losses demand to determine the electric power fed into the onshore grid in each operating range ($P_{t,r}^{grid}$) independently (columns in figure 1).

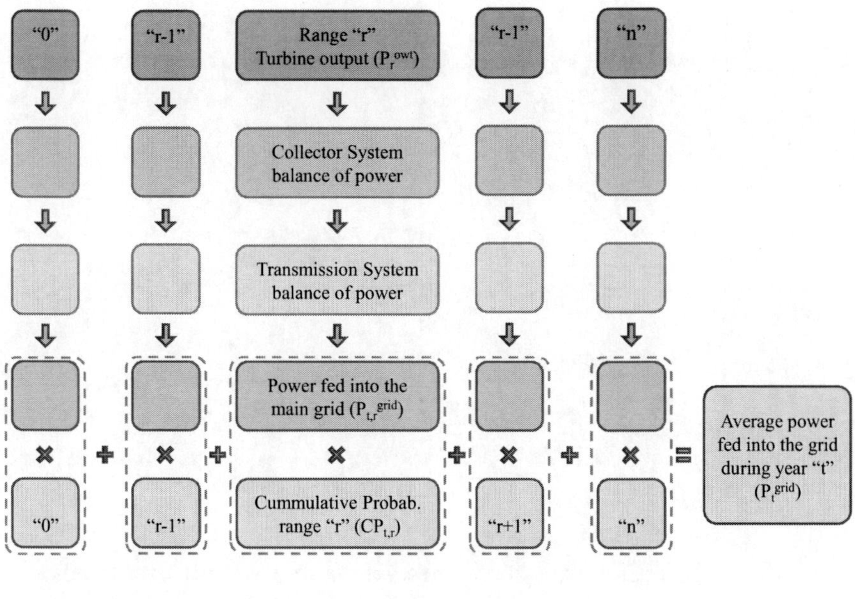

Figure 1. Assessment of the average power fed into the grid during year t.

Afterwards, the average power delivered to the grid during a year (P_t^{grid}) is obtained weighting the power delivered in each range by the cumulative probabilities ($CP_{t,r}$) for the wind blowing at each wind speed range. These probabilities are obtained using (3), which considers a Rayleigh probability density function with a mean wind speed at the site ($\overline{ws}$).

$$CP_{t,r}(ws) = e^{-\frac{(ws-0.5)^2}{2\sigma^2}} - e^{-\frac{(ws+0.5)^2}{2\sigma^2}}$$
$$\sigma = \sqrt{\frac{2}{\pi}}\,\overline{ws} \tag{3}$$

The power delivered to the grid in each operating range is determined as follows (each column in figure 1). First, the power curve of the OWTs is divided into steps and the average power output of the turbine in each range (P_r^{owt}) is taken as representative for the whole range (*e.g.* 1.63 MW between 8 and 9 m/s in figure 2).

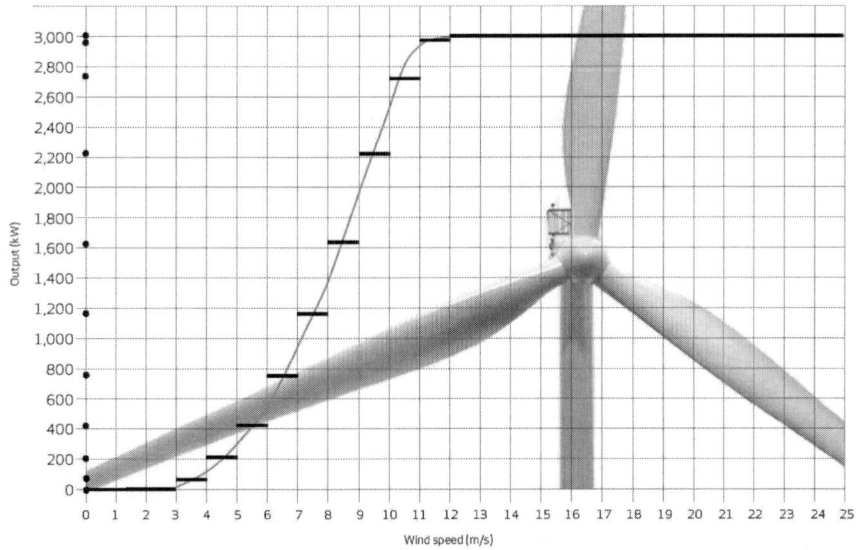

Figure 2. Power curve (Vestas) *binnarized* into 1 m/s steps.

From there on, a balance of power is performed considering the losses in all the components of the collector system (CS) and the transmission system (TS). The wake effect that lessens the energy captured by the OWTs and the unavailability of the different systems are also taken into account. The overall balance of power of the OWPP is set out in (4).

$$\begin{aligned}
P_{in}^{owpp} &= P_{in}^{cs} \\
P_{lost}^{owpp} &= P_{wake}^{owts} + P_{unav}^{owts} + P_{lost}^{cs} + P_{lost}^{ts} \\
&\quad P_{lost}^{cs} = P_{lost}^{cs,cab} + P_{unav}^{cs} \\
&\quad P_{lost}^{ts} = P_{lost}^{ts,ofco} + P_{lost}^{ts,ofpts} + P_{lost}^{ts,cab} + \\
&\qquad\qquad + P_{lost}^{ts,onpt} + P_{lost}^{ts,onco} + P_{unav}^{ts} \\
P_{out}^{owpp} &= P_{in}^{owpp} - P_{lost}^{owpp} \\
P_{t,r}^{grid} &= P_{out}^{owpp} - P_{unav}^{grid}
\end{aligned} \tag{4}$$

In each operating range, the momentary electric power production ($P_{t,r}^{grid}$) and the electric power losses are calculated separately. The power losses considered for the OWTs and for the grid are only due to the wake effect and to unavailability issues, and they are assessed using (5).

$$\begin{aligned}
P_{wake}^{owts} &= N_{clu}\, N_{owt,clu}\, P_r^{owt}\, F_{wake}^{owts} \\
P_{unav}^{owts} &= N_{clu}\, N_{owt,clu}\, P_r^{owt}\, (1 - F_{wake}^{owts})\, F_{unav}^{owts} \\
P_{unav}^{grid} &= P_{out}^{owpp}\, F_{unav}^{grid}
\end{aligned} \tag{5}$$

The rest of the power losses included in (4), occur in the CS and the TS, and the proposed assessment formulation is described in the following two subsections.

4.1. Balance of Power in the Collector System

The design options currently considered for CSs are the string, the dendrite shaped and the double sided ring types. In this subsection, it is assumed that all the clusters within the OWPP have the same number of turbines and the same string design. It is done so because the currently favoured double sided ring design is operated in an open loop during normal operation, and the balance of power in this type of CS would be the same as for a radial design. It is also assumed that tapering the clusters is not recommended for logistic reasons during installation and for maintenance issues. Therefore, the cable section is the same in all cable stretches. If that is not the case, each cluster with a different configuration should be analysed independently, although the input data and the formulation are the same as those presented below.

The overall balance of power in the CS is summarized in (6), taking into account power losses in cables and losses due to overall CS unavailability.

$$
\begin{aligned}
P_{in}^{cs} &= N_{clu}\, P_{in}^{clu} \\
P_{lost}^{cs} &= P_{lost}^{cs,cab} + P_{unav}^{cs} \\
&\quad P_{lost}^{cs,cab} = N_{clu}\, P_{lost}^{cs,clu} \\
&\quad P_{unav}^{cs} = \left(P_{in}^{cs} - P_{lost}^{cs,cab}\right) F_{unav}^{cs} \\
P_{out}^{cs} &= P_{in}^{cs} - P_{lost}^{cs}
\end{aligned}
\tag{6}
$$

For a given operating range, the power lost in a cluster and its power output depend on the number of OWTs feeding into it and on the characteristics of the submarine cables. The losses in MVAC cables have to be evaluated in each stretch separately and, afterwards, add them up to calculate the total power lost in the cluster. This is because the current through the inter-turbine cable increases as more turbines feed their power into the cluster.

Therefore, the balance of power in a cluster starts with the stretch farthest away from the offshore substation (OS), which is only fed by the last turbine in the cluster. Having calculated the electric power lost in the first stretch, the power fed into the second stretch is the output of the first stretch plus the power fed in by the second turbine in the cluster. And so on until the last MVAC cable stretch, going from the last turbine in the cluster to the MV switchgear placed at the OS. The balance of power in a cluster is presented in (7) and the technique proposed to assess the electric power losses in each cable stretch ($P_{lost}^{cs,str}$) is detailed in [26].

$$
\begin{aligned}
P_{in}^{cs,clu} &= N_{owt,clu}\, P_{r}^{owt}\left(1 - F_{wake}^{owt}\right)\left(1 - F_{unav}^{owt}\right) \\
P_{lost}^{cs,clu} &= \sum_{i=1}^{N_{owt,clu}} \left[P_{lost}^{cs,str}\right]_i
\end{aligned}
\tag{7}
$$

4.2. Balance of Power in the Transmission System

Throughout this subsection, a TS with the following characteristics has been considered to improve the readability of the paper: one offshore substation with two power transformers on it, a single HVAC three-core submarine cable and a single power transformer at the connection point substation (CPS). However, the method could be similarly applied to other scenarios, *e.g.* more than one OS in the windfarm, more than one cable per OS, single three-winding power transformer per OS, etc.

Five subsystems are taken into account in the balance of power of the transmission system: reactive power compensation equipment both at the OS and the CPS, power transformers both at the OS and the CPS and HVAC three-core cross linked polyethylene (XLPE) submarine cable. The overall balance of power in the TS described in (8) also includes a sixth item for TS unavailability losses.

$$\begin{aligned} P_{in}^{ts} &= P_{out}^{cs} \\ P_{lost}^{ts} &= P_{lost}^{ts,ofco} + P_{lost}^{ts,ofpts} + P_{lost}^{ts,cab} + \\ &\quad + P_{lost}^{ts,onpt} + P_{lost}^{ts,onco} + P_{unav}^{ts} \\ P_{out}^{ts} &= P_{in}^{ts} - P_{lost}^{ts} = P_{out}^{owpp} \end{aligned} \tag{8}$$

4.2.1. Offshore Reactive Power Compensation Equipment

Shunt fixed reactors are usually connected to the MVAC busbar at the OS. The losses in these reactors are assessed using (9). As we see, these losses vary with the square of the voltage, which will probably oscillate around the rated voltage $\pm$ 5% in normal operation, with a maximum of $\pm$ 10% at the CPS. For balance of power purposes, the losses in these reactors are considered the same for all the operating ranges, assuming that the voltage will oscillate up and down the rated value due to the varying nature of the power infeed.

$$P_{lost}^{ts,ofco} = 3\, I_{reac}^2\, R_{reac} \tag{9}$$

Where:

- $I_{reac} = U_{cs} \,/\, (\sqrt{3}\, Z_{reac})$
- $Z_{reac} = \sqrt{R_{reac}^2 + X_{reac}^2}$
- $R_{reac} = X_{reac} \,/\, Q$
- $X_{reac} = X_{reac}^{hv} \,/\, r^2$

- $r = U_{ts} \,/\, U_{cs}$
- $X_{reac}^{hv} = 2\, X_{line}$
- $X_{line} = X_L - X_C$
- $X_L = \omega\, L\, L_{ts,cab}$
- $X_C = (\omega\, C\, L_{ts,cab})^{-1}$

4.2.2. Offshore Power Transformers

It is usual to install two power transformers rated at 60-70 % of the power collected in the OS. The losses in the offshore power transformers, in the operating range under consideration, are calculated using (10), whereas iron and copper losses (P_{Fe}, P_{Cu}) are taken from figure 3 [27].

$$P_{lost}^{ts,ofpts} = 2\left[P_{Fe} + P_{Cu}\left(\frac{P_{in}^{ts,ofpts} \,/\, PF_{ts}}{S_{rated}^{ts,ofpts}}\right)^2\right] \tag{10}$$

Where:

- $S_{rated}^{ts,ofpts} = 0.6\,(P_{rated}^{owpp} \,/\, FP_{cs})$
- $P_{in}^{ts,ofpts} = 0.5\,(P_{in}^{ts} - P_{lost}^{ts,ofco})$

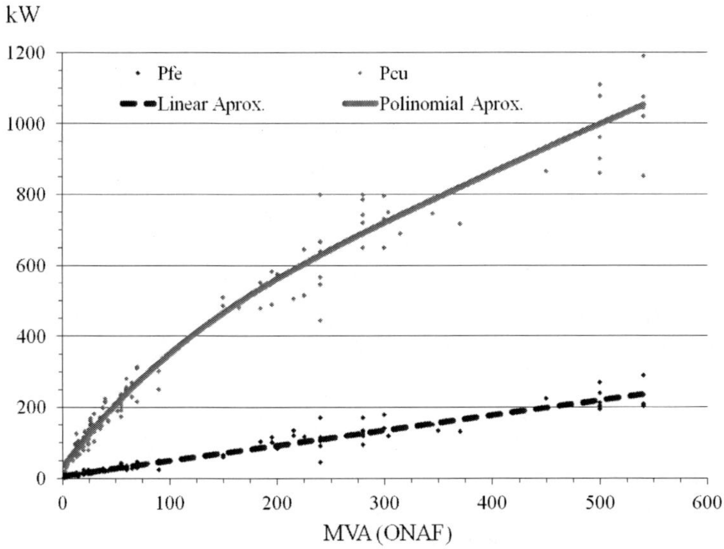

Figure 3. Iron and copper losses in oil insulated transformers [27].

4.2.3. HVAC Submarine Cable

The current in long HVAC submarine cables installed in OWPPs varies along its length mainly due to two phenomena [28, 29]. On the one hand, the intrinsic capacitive nature of these cables causes a variable charging current along the transmission line. On the other hand, the varying nature of the power in-feed from the OWPP causes strong variations in the current related to the active power transmission. This is why it is common practice to assess those losses with power flow simulation packages in industry and academia. Alternatively, [26] proposes a technique that allows to evaluate electric power losses in long HVAC submarine cables with enough accuracy and much faster. This new technique is set out in this paper within the overall method to assess techno-economically electric layouts of OWPPs.

4.2.4. Onshore Power Transformers

As commented on before an electric topology with a single full power rated transformer at the onshore CPS has been chosen to describe the formulation, with the losses assessed using (10) and the iron and copper losses obtained from figure 3. This is based in the fact that the solutions commonly chosen in the onshore substations of operational OWPPs have been either one three-phase full power rated transformer or a bank of three single-phase transformers. Other options such as two half rated power transformers or even two full rated power transformers have also been considered due to availability issues.

4.2.5. Onshore Reactive Power Compensation Equipment

Static synchronous compensators (STATCOMs) based on voltage source converter (VSC) technology are being planned and installed at the moment, in some cases in conjunction with fixed shunt reactors. For losses assessment purposes we will consider a STATCOM rated to half the reactive power of the HVAC cable (U_{ts}^2 / X_{react}^{hv}) and composed of a step-down power transformer and a multilevel modular converter (MMC). The losses in the transformer are evaluated as has been observed previously and the losses in the MMC are considered to be near 1.0 % at rated power [30, 31]. As in the case of the offshore reactive power compensation equipment, these losses are considered the same for all operating ranges.

4.2.6. Transmission System Unavailability

The power lost due to the unavailability of the TS is evaluated using (11). The electric power losses are discounted from the power coming into the TS to get a more accurate assessment of the power really lost due to forced outages of the transmission link. It makes sense to note that a similar strategy has been followed in (5) and (6) for the unavailabilities of the grid and the CS.

$$P_{unav}^{ts} = F_{unav}^{ts} \left[P_{in}^{ts} - \left(P_{lost}^{ts,ofco} + P_{lost}^{ts,ofpts} + + P_{lost}^{ts,cab} + P_{lost}^{ts,onpt} + P_{lost}^{ts,onco} \right) \right] \tag{11}$$

5. Validation

Before commenting on the case studies prepared to check different aspects of the method, it makes sense to point out some general aspects of the offshore wind business that prevent the validation of the method as a whole.

On the one hand, the results reported in most of the studies presented during the last decade (Ackermann, Brakelmann, Breseti, Lundberg, Walling) differ significantly from case to case, and sometimes the considered technical characteristics are not commented on in detail, which makes impossible to repeat the case studies faithfully [4]. Moreover, these results are not contrasted with actual data from offshore wind farms and since some of the studies were made years ago, the case studies would not be repeated today in the same way (different technical solutions and/or evaluation criteria).

On the other hand, actual information of commissioned offshore wind farms is currently available only for a few companies (Vattenfall, Dong, E.On, SSE, etc.). And they surely will not be keen on disclosing it because the advantages they have over newcomers rely, to some extent, on not repeating the errors made during the first projects. As information gets out of date, either due to the inclusion of new technical solutions or for substantial variations in costs, it is probably passed on to leading consultants (BGV Associates, Douglas Westwood, Garrad Hassan, Parsons Brinckerhoff, TNEI, etc.). Manufacturers and contractors involved in the first projects (Siemens, Vestas, Alstom, ABB, Prysmian, JDR, Fluor, etc.) surely have sound information, but only partial in most cases.

Unfortunately, at the time of writing this paper, these sources of information are not available, and therefore, the validation approach has been to check all the analytic calculations involved in the assessment of the yield of the OWPP, with special attention on the electric power losses assessment. Four issues have been checked independently:

1. Power losses in XLPE cables at rated power;
2. Average electric power losses in MVAC cables;
3. Average electric power losses in HVAC cables;
4. OWPP energy yield and overall electric power losses.

The results of the first three checks listed above as well as the formulation proposed to assess the electric power losses in XLPE three-core cables are detailed in [26]. The results show that the accuracy of the estimations is good enough for its purpose within the techno-economic assessment method.

As regards of the fourth check, the annual average electric power losses for different OWPP layouts have been evaluated with the method presented in this paper and considering the loss rate per AC/DC converter station close to that of the PWM based *classic* VSC technology. The topologies considered in the study for a 400 MW OWPP have been the small AC, small DC, large AC and AC/DC windfarms described in [5]. Figure 4 presents the annual average losses of each layout as a function of the transmission distance [3].

According to these results, the break even distance between HVAC and *classic* VSC-HVDC technologies for a 400 MW OWPP is near 85 km attending to the electric power losses. This result is somewhere between the 80 km reported in [10] and the 115 km reported in [32]. The results presented in [9] and [5] reported break even distances of 80 km and 130 km respectively, although these analyses were based on economic comparisons and not only on the electric power losses in the alternatives. These results validate the overall accuracy of the proposed method to some extent, because the assessment is influenced by all the steps proposed to assess the energy yield of an OWPP.

[3]Technology for small DC solutions is still in its infancy and far away from the industrial market, and proposing an AC/DC solution for a 400 MW OWPP makes no sense in the current state of the market. It is therefore worth it to point out that the objective of evaluating these layouts is to apply the proposed method to the same scenarios analysed in previous studies, compare the results, and to a certain extent, validate the accuracy of the proposed method.

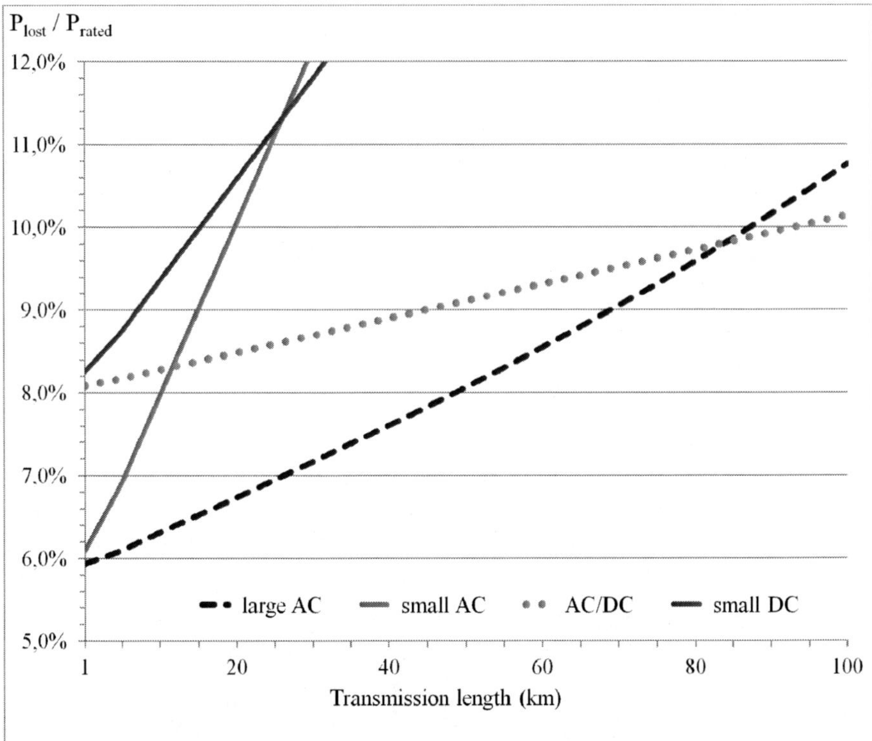

Figure 4. OWPP overall losses assessment check.

6. Application of the Methodology

6.1. Scope of Application

The proposed methodology can be helpful in evaluating the performance of alternative electric topologies[4] mainly in two scenarios:

- The optioneering studies conducted in the early stages of pre-FEED design in industrial offshore wind projects;
- The assessment of the economic potential of novel solutions under consideration within applied academic research.

[4]The electric topology or layout of an OWPP is understood here as an unique combination of OWTs, MV cables, offshore substations (OSs), HV cables and CPS with all technical variables fixed, *e.g.* wind energy conversion system (WECS) type and rated power, CS voltage, number and type of OSs and transformers, TS voltage, number of cables, CSAs, etc.

in the optioneering studies of offshore wind projects, or in assessing the projection of novel technical solutions under research.

For the former, the technologist could determine the NPV of each alternative to choose afterwards the solution that leads to a higher amount. For the latter, the researcher might evaluate a realistic project in two situations: one including the innovation under study and another without it. The difference between the NPVs obtained can be taken as an indicator of the potential interest of his proposal. In order to obtain reliable conclusions from the analysis, it is up to the researcher to evaluate the effect of the new solution in terms of project variables (CAPEX, energy yield or OPEX).

In relation to the case studies realistically approachable, although the formulation has been detailed for pure AC windfarms it can easily be expanded to AC/DC windfarms. The losses in HVDC cables are easier to assess than those of HVAC cables, and an approach to characterize VSC converter losses at system level has also been shown. However, the main difficulty nowadays relies on getting realistic input data for the CAPEX and OPEX of the alternatives evaluated, mainly for OWPPs with MMC based VSC-HVDC links to shore[5].

This difficulty affects researchers and OWPP developers in a different way. Researchers have the advantage of not being so interested in the actual NPV of the project, as they are in evaluating the projection of the innovation they are considering. However, as HVDC manufacturers do not usually disclose that kind of information if they are not bidding, researchers will probably have to be content with small pieces of information scattered about in papers and reports.

OWPP developers, and the consultants to which they order the grid integration studies, have the advantage of having at their disposal first hand information from manufacturers and contractors. However, their evaluation will be taken as a reference for the key decisions during project optioneering, and this is a huge responsibility. In both cases, this difficulty will persist until experience is gained in on-going projects and some information is afterwards disclosed.

6.2. An Example: OWT Performance Assessment

This subsection illustrates how to apply the proposed methodology to compare alternative OWTs for a project. The example considers two turbines currently

[5]The Offshore Renewable Energy Catapult (`https://ore.catapult.org.uk/`) is currently working with industrial and academic partners to address this challenge. The results of an ongoing study focused in this issue are expected to be published later on this year.

available in the market, and the characteristics of the project have been chosen according to a real case. The presentation of the case study is structured into three steps:

1. Prepare technical and economical input data.
2. Shape the formulation in a spreadsheet or algorithm.
3. Perform the calculation and interpret the results.

6.2.1. Prepare Technical and Economical Input Data

The following list summarizes the technical characteristic of the OWPP with two alternative OWTs. In the case of input data with different values in both alternatives they are explicitly indicated. Otherwise data is given once.

- A 400 MW OWPP with 80 OWTs.
- Site with a mean wind speed of 9.78 m/s at 100 m high, and lifetime variations following normal distribution with 10 % standard deviation.
- An average lowest astronomical tide (LAT) within the OWPP of 19 m, at a site with a mean high water spring (MHWS) of 8.3 m.
- Two OWTs with a rated power of 5 MW, 690 V generators and 0.690/33 kV power transformers:
  - OWT-1: DFIG WECS, cut-in at 4 m/s, rated power at 13 m/s, cut-out at 30 m/s, rotor diameter 126 m.
  - OWT-2: PMSG WECS, cut-in at 2 m/s, rated power at 11 m/s, cut-out at 28 m/s, rotor diameter 128 m.
- 8 clusters of 10 turbines with a double sided ring design.
- 33 kV XLPE three-core 185 mm^2 cables for the CS.
- Cross-wind distance among OWTs 6 times rotor diameter.
- Two OSs with two 33/132 kV transformers on each, rated at 60 % of the power fed into the OS by four clusters.
- Fixed shunt reactors with a inductive reactance double the reactance of the HVAC cable connected to each OS.
- One 132 kV XLPE three-core 800 mm^2 cable per OS.
- 35 km between the OSs and the landfall and 1.5 km between the landfall and the CPS.

- Two static synchronous compensator (STATCOM) at the CPS rated at half the reactive power of a HVAC cable.
- One 132/400 kV power transformer at the CPS.

The list below presents the values fixed for a number of variables and parameters. They have been chosen considering several books and reports [33, 34, 35, 20, 21, 22, 23, 4, 36, 37, 38].

- 15 % reduction in OWT production due to the wake effect (10-20 %).
- OWTs unavailability (5-15 %):
  - OWT-1: 8.0 % (turbine with operational experience in commissioned OWFs).
  - OWT-2: 10.0 % (new turbine in test period).
- CS unavailability: 1.5 %.
- TS unavailability: 2.5 %.
- Grid unavailability: 1.0 %.
- Lifetime: 20 years (20-25).
- CAPEX (2.8-3.3 M£/MW):
  - OWT-1: 1246 M£
  - OWT-2: 1323 M£ (improved performance at expenses of a 15 % higher cost)
- Loan years: 12.
- Interest rate: 7 %.
- OPEX: 32 M£/year during 5 years and 2.5 % reduction from there on (0.4-3.0 % CAPEX/year).
- Electricity price: 120.54 £/MWh for the first five years with 7 % annual depreciation for the rest.
- Corporate tax rate: 26 %.
- Discount rate: 10 %.

6.2.2. Shape the Formulation in a Spreadsheet or Algorithm

Figure 5 presents what (1) and (2) could look like shaped in a spreadsheet. The NPV shown in the 10th row is calculated applying (1) to the net cash flows in the

9th row (the discount factor and corporate tax rate are not explicitly shown). The series of annual electricity prices and OPEX are fixed according to the information provided above (2nd and 6th rows). Regarding the wind speed, completely different series could be proposed with a mean value of 9.78 m/s and a 10 % standard deviation. The 3rd row shows one of these feasible series intended to highlight one of the virtues of the proposed method: it enables to assess effectively the influence of long term variations of different project variables in conjunction (wind speed, electricity price, etc.). The commercial loan installments (7th row) are calculated at constant interest rate and fees due at the end of each year, whereas the incomes provided by the energy sold yearly (5th row) are assessed with (2).

Year	1	2	3	4	5		12	13	14	15	16	17	18	19	20
Elec. price (£/MWh)	120,54	120,54	120,54	120,54	120,54		72,53	67,45	62,73	58,34	54,26	50,46	46,92	43,64	40,59
Wind speed (m/s)	9,78	9,00	9,58	8,31	10,27		8,80	10,17	10,76	10,07	10,95	10,17	9,39	9,49	10,27
Energy yield (TWh)	1.361	1.231	1.329	1.101	1.433		1.194	1.419	1.498	1.404	1.522	1.419	1.298	1.314	1.433
Income (M£)	164,0	148,3	160,2	132,7	172,8		86,6	95,7	94,0	81,9	82,6	71,6	60,9	57,4	58,2
OPEX (M£)	-32,0	-32,0	-32,0	-32,0	-32,0		-26,8	-26,1	-25,5	-24,8	-24,2	-23,6	-23,0	-22,4	-21,9
Installment (M£)	-156,9	-156,9	-156,9	-156,9	-156,9		-156,9								
Cash Flow (M£)	-24,9	-40,5	-28,7	-56,2	-16,1		-97,1	69,6	68,5	57,1	58,4	48,0	37,9	34,9	36,3
After Tax CF (M£)	-24,9	-40,5	-28,7	-56,2	-16,1		-97,1	51,5	50,7	42,2	43,2	35,5	28,0	25,8	26,8

NPV (M£)	-243,0

Figure 5. Project net present value assessment in a spreadsheet.

Figure 6 is intended to give an idea of how to shape equations (3) to (11) in a spreadsheet. This calculation has to be repeated for each year of the lifetime. However, once the spreadsheet is shaped, just the mean wind speed needs to be changed to assess the average electric power feed into the main grid during the year under evaluation.

The power curve of the turbine is binnarized in 1 m/s steps (rows with blue background). Given a mean wind speed of 9,78 m/s, the cumulative probabilities for the wind blowing at each speed range are obtained using (3) (rows with green background).

For each speed range, the power that the OWPP effectively feeds into the main grid is assessed independently. For example, with the wind blowing at 10 m/s each turbine provides 3333 kW. From there on, a balance of power is performed for the CS and another one for the TS. These balances of power (violet and orange backgrounds in the figure) consider equations (4) to (11).

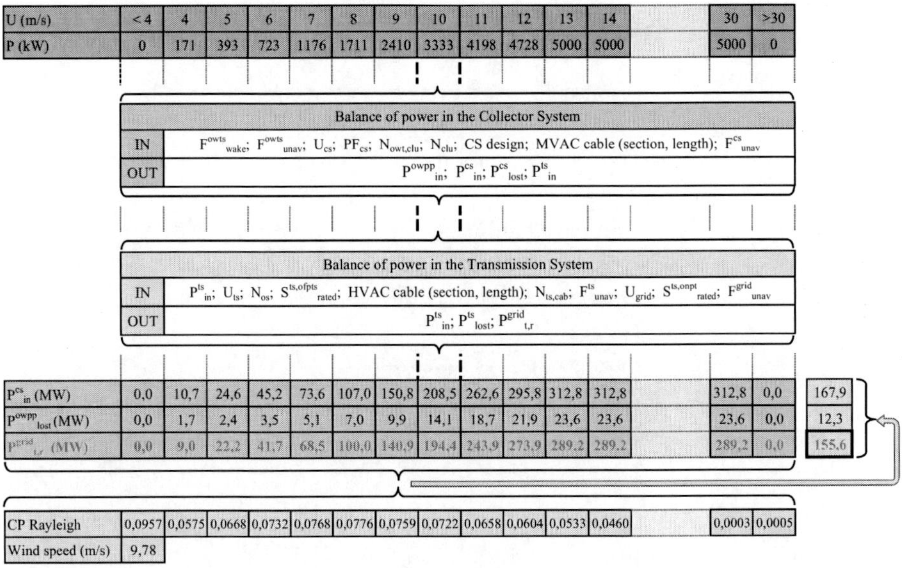

Figure 6. Annual average power production assessment in a spreadsheet.

The output of both balances of power are the assessed momentary values for all the magnitudes considered in (4). Figure 6 only shows three of these magnitudes in the rows with a red background:

- 208.5 MW feed by all the OWTs into the CS (P_{in}^{cs}).
- 14.1 MW lost in the CS and TS (P_{owpp}^{lost}).
- 194.4 MW feed into the onshore grid ($P_{grid}^{t,r}$).

The process is repeated for all the operating ranges, which is quite simple to do in a spreadsheet once the formulation is shaped for one of them. The momentary power values (red rows) are averaged with the cumulative probabilities of each range (green row) to obtain annual average power values (cinnamon box). For example, with OWT-1 an average of 155.6 MW are effectively fed into the grid during a year with a mean wind speed of 9.78 m/s. The income for the purchased energy can then be easily assessed with (2).

The average values for all the power magnitudes included in (4) are obtained similarly and they also constitute a very valuable output of the method. The lifetime losses breakdown for the case study with OWT-2 is shown in table 1.

Table 1. OWPP lifetime losses breakdown

Losses	TWh	(%)
OWTs unavailability	2540	54.0
MVAC cables	162	3.4
CS unavailability	437	9.3
Offshore shunt reactors	37	0.8
Offshore power transformers	89	1.9
HVAC cables	318	6.8
Onshore power transformer	73	1.6
Onshore STATCOM	65	1.4
TS unavailability	699	14.9
Grid unavailability	280	5.9
	4700	100.0

6.2.3. Interpret the Results

The result of the assessment is a negative NPV of 243.0 M£ for OWT-1 (figure 5) and a positive NPV of 6.3 M£ for OWT-2. Therefore, the PMSG based turbine is clearly preferred. It is worth noting that the NPV of the project is negative with OWT-1 and only slightly positive with OWT-2. Some reasons for this can be found analysing of the sets of input data chosen:

1. The proposed annual average wind speeds (3rd row in figure 5) are well below the nominal average of the site during the first years, when the electricity prices are at their highest. The negative effect that this has on incomes, is not compensated by the good wind speeds during the last few years. This is due to the lower electricity price during the last years of the project, and to the time value of money. This result demonstrates the relevance of considering the lifetime variation of the wind speed and the electricity price in conjunction in order to assess the risks during the feasibility stage of these projects. It also highlights the possibility offered by the new methodology to evaluate these scenarios in an agile way.

2. The wake effect and unavailability factors considered carry a significant reduction in the estimated production of the windfarm. Overall OWPP

unavailabilities around 12 % are considered in some of the cited references which indicates that the assessment presented is realistic at the moment. However, the availability of all the subsystems of the OWPP is expected to increase as more experience is gained by the industry.

3. The 10 % discount factor has been widely used in this kind of analyses during the last few years, assuming funding schemes based on private equities or venture capitals. On the contrary, the proposed methodology considers an interest rate to be paid to the public investors, and therefore, it seems reasonable to consider a lower discount factor in the assessment of the time value of money. However, deeper economic analyses are outside the scope of the paper.

Finally, another issue that arouses interest is the actual breakdown of the energy lost in the OWPP during its lifetime. According to the results shown in table 1, it is clear that improving the availability of the different systems that make up the windfarm is one of the main objectives for companies and researchers committed with offshore wind development.

Conclusion

A new method to assess techno-economically innovative solutions for offshore wind has been presented. It can be applied to the evaluation of new ideas in the context of applied engineering research within academic and/or industrial spaces. It can also be applied to the evaluation and comparison of different electric layouts during the early stages of pre-FEED offshore wind power plant design. The list below summarizes its main advantages when compared with previous approaches:

1. It performs a life-cycle economic analysis taking into account the time value of money, oncoming financing schemes funded on public equity markets and other issues such as the wake effect and the unavailability of the systems that make up the OWPP.

2. It can be shaped into a spreadsheet or in algorithmic form avoiding the need of power flow simulation. Therefore, the time needed to develop the studies is decreased and the number of researchers, engineers and technologists capable of conducting the assessments is increased.

3. It allows the agile evaluation of the correlated influence of long term variations in key variables such as wind speed, electricity prices, OPEX, etc. This feature of the new methodology could make it suitable to support quantitatively the risk assessments usually conducted during the early stages of offshore wind projects.

References

[1] " Offshore Wind Cost Reduction: Pathways Study," The Crown Estate, *Tech. Rep.*, May 2012.

[2] "Offshore Wind Cost Reduction Task Force Report," Department of Energy and Climate Change (DECC) of the UK, *Tech. Rep.*, June 2012.

[3] "Economic Impact Study into the Development of the UK Offshore Renewable Energy Industry to 2020," Offshore Renewable Energy Catapult, *Tech. Rep.*, March 2014.

[4] A. Madariaga, I. Zamora, J. L. Martín, I. Martínez de Alegría, and S. Ceballos, *Evaluation methodologies and input data for offshore wind power plant topology assessment.* NOVA Science Publishers, Inc., New York, 2013, vol. Renewable Energy: Economics, Emerging Technologies and Global Practices, ch. 7, pp. 111–160.

[5] S. Lundberg, "Performance comparison of wind park configurations," Deptartment of Electric Power Engineering, Chalmers University of Technology, *Tech. Rep.*, 2003.

[6] T. Ackermann, N. Barberis Negra, J. Todorovic, and L. Lazaridis, "Evaluation of electrical transmission concepts for large offshore wind farms," in *Copenhagen Offshore Wind*, October 2005.

[7] L. Lazaridis, "Economic comparison of HVAC and HVDC solutions for large offshore wind farms under special consideration of reliability," *Master's thesis*, Royal Institute of Technology, Department of Electrical Engineering, Stockholm, Sweden, 2005.

[8] L. Fingersh, M. Hand, and A. Laxson, "Wind turbine design cost and scaling model," National Renewable Energy Laboratory of the U.S. (NREL), *Tech. Rep.* NREL/TP-500-40566, December 2006.

[9] P. Bresesti, W. L. Kling, R. L. Hendriks, and R. Vailati, "HVDC connection of offshore wind farms to the transmission system," *IEEE Transactions on Energy Conversion*, vol. 22, no. 1, pp. 37–43, 2007.

[10] B. Van Eeckhout, D. Van Hertem, M. Reza, K. Srivastava, and R. Belmans, "Economic comparison of VSC HVDC and HVAC as transmission system for a 300 MW offshore wind farm," *European Transactions On Electrical Power*, vol. 20, no. 5, pp. 661–671, 2010.

[11] N. B. Negra, J. Todorovic, and T. Ackermann, "Loss evaluation of HVAC and HVDC transmission solutions for large offshore wind farms," *Electric Power Systems Research*, vol. 76, no. 11, pp. 916–927, 2006.

[12] G. Quiñonez Varela, G. W. Ault, O. Anaya-Lara, and J. R. McDonald, "Electrical collector system options for large offshore wind farms," *IET Renewable Power Generation*, vol. 1, no. 2, pp. 107–114, 2007.

[13] R. A. Walling and T. Ruddy, "Economic optimization of offshore substations and collection systems," in *Proceedings of the 5th International Workshop on Large-Scale Integration of Wind Power and Transmission Networks for Offshore Wind Farms*, Glasgow, Scotland, April 2005.

[14] M. Scutariu, "Techno-economical optioneering of offshore wind farms electrical systems," in *Proceedings of the 7th conference of the IEEE Power & Energy Society in Europe (PowerTech)*, Lausane, Switzerland, July 2007.

[15] E. Camm, M. Behnke, O. Bolado, M. Bollen, M. Bradt, C. Brooks, W. Dilling, M. Edds, W. Hejdak, D. Houseman, S. Klein, F. Li, J. Li, P. Maibach, T. Nicolai, J. Patino, S. Pasupulati, N. Samaan, S. Saylors, T. Siebert, T. Smith, M. Starke, and R. Walling, "Wind Power Plant Substation and Collector System Redundancy, Reliability, and Economics (IEEE PES Wind Plant Collector System Design Working Group)," in *Proceedings of the IEEE Power & Energy Society (PES) General Meeting*, Calgary, Canada, July 2009.

[16] M. Zhao, Z. Chen, and F. Blaabjerg, "Optimisation of electrical system for offshore wind farms via genetic algorithm," *IET Renewable Power Generation*, vol. 3, no. 2, pp. 205–216, 2009.

[17] J. S. González, A. G. G. Rodríguez, J. C. Mora, J. R. Santos, and M. B. Payan, "Optimization of wind farm turbines layout using an evolutive algorithm," *Renewable Energy*, vol. 35, no. 8, pp. 1671–1681, 2010.

[18] M. Banzo and A. Ramos, "Stochastic optimization model for electric power system planning of offshore wind farms," *IEEE Transaction on Power Systems*, vol. 26, no. 3, pp. 1338–1348, 2011.

[19] P. Pezzini, O. Gomis-Bellmunt, and A. Sudri-Andreu, "Optimization techniques to improve energy efficiency in power systems," *Renewable and Sustainable Energy Reviews*, vol. 15, no. 4, pp. 2028–2041, 2011.

[20] "Offshore Wind: Forecasts of future costs and benefits," BGVAssociates in behalf of RenewableUK, *Tech. Rep.*, June 2011.

[21] "Who's winning the clean energy race?" The PEW Charitable Trusts, *Tech. Rep.*, April 2012.

[22] "A Guide to Offshore Wind and Oil & Gas Capability," Scottish Enterprise, *Tech. Rep.*, September 2011.

[23] "Review of the generation costs and deployment potential of renewable electricity technologies in the UK," U.K. Department of Energy and Climate Change, *Tech. Rep.*, October 2011.

[24] "A National Offshore Wind Strategy: Creating an Offshore Wind Industry in the United States," U.S. Department of Energy - Energy Efficiency & Renewable Energy, *Tech. Rep.*, February 2011.

[25] "State of the Industry Report. Onshore and Offshore Wind: A Progress Update," RenewableUK, *Tech. Rep.*, October 2011.

[26] A. Madariaga, J. L. Martín, I. Zamora, S. Ceballos, and O. Anaya-Lara, "Effective assessment of electric power losses in three-core XLPE cables," *IEEE Transactions on Power Systems*, vol. 28, no. 4, pp. 4488–4495, November 2013.

[27] (2013) SENER Ingeniería y Sistemas S.A. Electricity section. Getxo, Spain. http://www.sener.es.

[28] H. Brakelmann, "Efficiency of HVAC power transmission from offshore-windmills to the grid," in *Proceedings of the 5th conference of the IEEE Power & Energy Society in Europe (PowerTech)*, Bologna, Italy, June 2003.

[29] ——, "Loss determination for long three-phase high-voltage submarine cables," *European Transaction on Electric Power*, vol. 13, no. 3, pp. 193–197, 2003.

[30] J. Dorn, H. Gambach, J. Strauss, T. Westerweller, and J. Alligan, "Trans Bay Cable - A breakthrough of VSC multilevel converters in HVDC transmission," in *CIGRE Colloquium on HVDC and Power Electronic Systems, San Francisco, USA*, March 2012.

[31] M. Albertini, S. Franchi-Bononi, N. Kelly, M. Marelli, A. Miramonti, A. Orini, G. Perego, and G. Pozzati, "Innovation and applications for extruded HVDC cable systems," in *CIGRE Colloquium on HVDC and Power Electronic Systems, San Francisco, USA*, March 2012.

[32] S. Meier and P. C. Kjaer, "Benchmark of annual energy production for different wind farm topologies," in *Proceedings of the 36th Annual Power Electronic Specialists Conference (PESC)*, Recife, Brazil, June 2005.

[33] "Case Study: European Offshore Wind Farms - A Survey for the Analysis of the Experiences and Lessons Learnt by Developers," Deutsche energie agentur (DENA) and Deutsche WindGuard, *Tech. Rep.*, December 2007.

[34] G. Gaudiosi and J. Twidell, Eds., *Offshore Wind Power*. Multi-Science Publishing Co. Ltd, April 2009.

[35] "A guide to an offshore wind farm," BGV Associates on behalf of The Crown State, *Tech. Rep.*, January 2010.

[36] K. E. Thomsen, *Offshore Wind: A Comprehensive Guide to Successful Offshore Wind Farm Installation*. Elsevier Inc., 2012.

[37] (2012) *German feed-in tariffs for offshore wind farms.* http://www.germanenergyblog.de/?page_id=8617. Retrieved on May 2013.

[38] (2012) *Corporation Tax rates in the U.K. (HM Revenue & Customs).* http://www.hmrc.gov.uk/rates/corp.htm. Retrieved on May 2013.

INDEX

A

B

C

D

E

F

G

H

N

O

P

R

S

T

U

V

W

X

Y

Z